History, Philosophy and Theory of the Life Sciences

Volume 38

Alexandre Métraux, Archives Poincaré, Nancy, Germany
Hans Metz, Leiden University, Leiden, The Netherlands
Roberta L. Millstein, University of California, Davis, USA
Staffan Müller-Wille, University of Cambridge, Cambridge, UK
François Munoz, Université Montpellier 2, Montpellier, France
Dominic Murphy, University of Sydney, Camperdown, Australia
Stuart A. Newman, New York Medical College, Valhalla, USA
Frederik Nijhout, Duke University, Durham, USA
Samir Okasha, University of Bristol, Bristol, UK
Susan Oyama, The City University of New York, New York, USA
Kevin Padian, University of California, Berkeley, USA
David Queller, Washington University in St. Louis, St. Louis, USA
Stephane Schmitt, Archives Poincaré, Nancy, France
Phillip Sloan, University of Notre Dame, Notre Dame, USA
Jacqueline Sullivan, Western University, London, Canada
Giuseppe Testa, University of Milan, Milan, Italy
J. Scott Turner, SUNY College of Environmental Science and Forestry, Syracuse, USA
Denis Walsh, University of Toronto, Toronto, Canada
Marcel Weber, University of Geneva, Geneva, Switzerland

Guglielmo Militello

Functional Integration

A Theoretical Enquiry into the Biological Unit of the Individual

Springer

Guglielmo Militello
Institut d'histoire et de philosophie des sciences et des techniques
(IHPST, CNRS/Paris 1 Panthéon-Sorbonne)
Paris, France

ISSN 2211-1948 ISSN 2211-1956 (electronic)
History, Philosophy and Theory of the Life Sciences
ISBN 978-3-031-94743-8 ISBN 978-3-031-94744-5 (eBook)
https://doi.org/10.1007/978-3-031-94744-5

This work was supported by Universidad del Pais Vasco, Spain

This Springer imprint is published by the registered company Springer Nature Switzerland AG
The registered company address is: Gewerbestrasse 11, 6330 Cham, Switzerland

Dans ses écrits un sage Italien
Dit que le mieux est l'ennemi du bien
Voltaire, La Bégueule (1772)

To my teachers

Foreword

How many living beings are there on our planet? This question presupposes that we can clearly delimit organisms from their environment and distinguish them from others in order to detect, compare, and count them. We assume, of course, that (more or less) we can. Otherwise, how are we going to talk about evolution—and therefore, about re-production or species—without being able to individualize the immense diversity of life? And what about physiology, where we investigate how a living organism self-regulates, for example its body temperature, in response to environmental challenges?

These questions open the way to a more fundamental one, regarding whether and how we can provide principled ways to delimit and identify organisms rather than just rely on intuition. These individuated living entities called "organisms" are often elusive. Like social entities, biological organisms challenge our efforts at delimitation. It is not a case that "what is an individual?" is a central question in the philosophy of biology: the object of an intense and sophisticated debate involving evolutionary, physiological, and ecological considerations.

One way to look at biological individuality is to consider that the parts of living organisms are highly integrated. That is one important element that makes organisms cohesive entities that we can identify, track, count, and compare. What does integration mean? At first approximation, integration refers to how parts and their activities are kept together within a system. This notion is usually expressed in terms of mutual dependence between parts in a system in which these very parts perform different activities aimed at maintaining the system itself. In biological organisms, this means that: (1) the parts that play different roles are built, maintained, and repaired within the system; and (2) they are controlled and coordinated so that they can operate in such a way that they contribute to the overall functioning of the system as a cohesive whole.

There are very good reasons to reflect on these difficult but fundamental questions: this book is an excellent and original contribution to the advancement of our understanding of what makes a biological system an individual. As the title suggests, the main thesis of Guglielmo Militello's book is that the concept of functional integration is one or the core notion for understanding biological individuality.

Let us consider integration in a single cell: proteins are built when the cell needs them, and their activities are controlled to support essential cellular functions. These proteins catalyze biochemical reactions, operate as membrane channels, facilitate molecular transport throughout the cell, interact with nucleic acids, and modify the activities of other proteins, among other critical roles. During the history of life, unicellular organisms have come together at least 25 times to form cohesive multicellular associations in which different groups of cells are mutually dependent for their own existence and activity. Some of those are considered organisms as well.

This, again, is just part of a wider and more complex story. Living organisms have established between themselves a host of interrelations that have led to the appearance of new (temporal or permanent) composite entities. Some of them are based on loose and transient links. The partners can easily recover their independence, such as in the case of facultative multicellularity, where free living unicellular organisms come together in response to environmental cues, or in certain symbiotic associations like mycorrhizae between plants and fungi. Others, instead, are so strongly cohesive and have developed so deep and intense links that one or both partners can no longer exist as separated individuals. This is the case of obligated symbiosis such as between fungi and algae in lichens, between aphids and those endosymbiotic bacteria that provide essential nutrients that are absent from the host's diet, or the now endangered one between corals polyps and photosynthetic algae. In between these different examples there is an almost infinite number of intermediate cases.

In some sense, most of these composite entities could also be considered as individuals because they are characterized by mutually dependent parts, even though these "individuals" often cannot reproduce as a single unit and lack a common metabolism or a collective physical border. However, the study of these intermediate "individuals" raises challenging issues that lead to questioning shared assumptions about what biological entities are. Importantly, however, they also offer precious keys to understanding crucial evolutionary transitions, such as the origins of multicellular organisms or, as is the case of this book, the process of eukaryogenesis.

One might want to address those questions based on philosophical considerations and seek support in biological examples. Militello, instead, considering the complexity of the issue, takes the opposite direction and builds his philosophical work upon a solid and enlightening example. More, specifically, by looking at eukaryogenesis, his book deals with the question of how to characterize an organism as an individuated and cohesive entity: a system that cannot be divided into parts capable of independent life and that reproduces as a whole. It delves into how previously independent cellular units merged and underwent such dramatic reorganizations that they lost their independence and gave rise to a new cohesive system: the eukaryotic cell. At the center of all this is precisely the idea of "functional integration."

Though centered on the specific phenomenon, eukaryogenesis, the originality and main interest—and merit—of this book lies precisely in studying the problem of functional integration. By building upon a biological case study, the author offers a detailed analysis and precise definition of what he means by functional integration in biology. By focusing on such a central notion, he also contributes to developing, more generally, a framework to address the very question of biological individuality.

The concept of functional integration is widely used within and beyond biology. In neuroscience, for example, it is used to describe how brain regions work together to process information so as to generate a unified action (e.g., how the activities of large clusters of neurons are coordinated so as to fire together to generate a new brain function). In engineering, functional integration is understood as a design process that ends in a single new system capable of achieving multiple functions formerly realized separately in different systems. To achieve functional integration means to align a set of processes, systems, and resources so as to achieve an effective synergy, so that additional functions are fulfilled by the new system. In sociology, functional integration is understood as the process by which various parts of a society work together to maintain cohesiveness. It focuses on how different social institutions interact and contribute to the overall functioning of a society.

Although the concept of functional integration is so widely used that it seems at first glance easy to understand—integration as mutual dependence between things or the process of combining two or more things to create a whole—this apparent simplicity hides a complex and surprising phenomenon. But what is the core meaning of this concept? Suppose that several initially separated systems come in contact. Likely, this contact will alter the properties or activities of these systems and might produce conflicts between them. Perhaps some systems disintegrate during this process; or perhaps some of them are modified so that together they constitute a new and unique system, displaying new properties. This process—by which different independent systems begin to interact, modifying their parts and internal relations in such a way that, at the end, they become mutually dependent for their existence and activity and constitute a new, more complex and larger system—is the emergence of a new functionally integrated system.

In the literature on biological individuality, functional integration is a core concept. However, it is mostly adopted as an *explanans*. It assumed at the outset and used to describe phenomena by which things come together or to discuss the notion of individuality. It is not itself the target of investigation. This book adopts an innovative approach and takes it as an *explanandum*. It tackles it directly, analyzes it, discusses its uses, and develops an account of it through the lens of a specific, yet paradigmatic, biological phenomenon.

A specific merit of this book is to choose a domain—eukaryogenesis—that makes it possible to address all these issues, offering two main advantages. Eukaryogenesis provides the lens through which Militello chooses to address those questions revolving around the problem of biological individuality. Ontologically, the phenomenon of eukaryogenesis is one of the earlier forms in which functional integration appears in Nature. In this respect, it has similarities with the origins of life as a case in which a new functionally integrated biological regime is established. However, compared to the origins of life, eukaryogenesis has the advantage that it is a well-studied domain, with abundance of data. At the same time, it is characterized by a fertile debate, with different competing theories. Hence, the choice of this domain is a sensible research strategy.

Now, behind this focus on functional integration through the lens of eukaryogenesis, there is another remarkable merit of this book. Unlike the common views on organisms as biochemical machines driven by a complex genetic structure, and of biological individuals characterized as units of selection, Militello's approach to individuality is organizational. What is "an organizational approach"? Generally, the organization of a system is defined as the way how its different parts (each performing different causal roles) interact in specific, coordinated ways so as to collectively support some specific global/collective operation. An organization is not merely an abstract set of relations between parts, but a (thermo)dynamic process, where these parts perform mutually dependent local actions, each of which contributes to the global operation.

To realize such network of activities, parts need to satisfy many material requirements: they need to have some given shapes, sizes, and material compositions. Human-made organizations such as a machine can have very resistant parts, which shapes and properties can remain unchanged over a long time. Instead, a biological organization relies on molecular components, such as proteins, which are highly dynamic, have a short lifespan and are constantly transformed and degraded. Thus, unlike those found in human artifacts, these components need to be continuously replaced or repaired. To do so an organism uses these components to procure and take in resources (matter and energy), process them, and use them to select and run their internal processes in ways that are appropriate to circumstances, to build and repair their components, and to generate and modulate behaviors that allow them to move in and interact with their environment. Although not all biological organizations are equally integrated and cohesive, the different parts and processes are highly interdependent. What is common is the need to coordinate the activities of the components in space and time and manage their different energetic needs, so that they not only avoid conflict but can mutually contribute the maintenance of the organism.

Work on functional integration fits perfectly within this perspective. The different ways of realizing functional integration that Militello analyzes in his book make full sense within an organizational perspective. Let us just think of the many trade-offs between different energetic and physiological processes necessary to maintain an integrated system, or the evolutionary bottlenecks systems had to overcome to establish new forms of integration. They are carefully analyzed in the book, showing that different ways biological systems are individuated or become individuated can be explained in terms of how they are actually organized. When biological components or systems closely interact, conflicts may appear, as different parts may have different requirements, often not compatible or even mutually exclusive. Integration is achieved by employing regulatory controls to coordinate the activities of these parts. These include mechanisms for the control of the relationship with the environment, because the viability of any system whose identity is based on a logic of self-construction depends on the specific energy requests and the actual rates of their always precarious dynamics. This integration can be achieved in many different ways. Between strongly integrated systems (for example, a paradigmatic example of an organism such as a bacterial cell) and a loosely distributed network (a colony), there is an interesting domain of more or less integrated forms of organization. Analyzing different ways of realizing functional integration can therefore shed light on the phenomenon of individuation and give a crucial contribution to the debate on biological individuality.

Last, but not least, we would like to highlight another non-common merit of this book: the fruitful way in which Militello integrates philosophical analysis with scientific study. He does so by borrowing updated data and theories from biology, and revising these ideas through a philosophical analysis, in such a way as to put the two perspectives into dialogue. Through this mutual enriching interaction between biology and philosophy as it unfolds in this book, he develops new ideas that can help to orient future scientific research in the understanding of the phenomenon of functional integration and the role it has played in the evolution of life. At the same time, this approach is characterized by specific methodological commitments. The book is written from a philosophical perspective, but Militello keeps close to empirical knowledge and scientifically grounded theories, avoiding a priori speculations and formulating theoretical proposals capable of being testable by empirical research. In this way, philosophical research could—and should—be ultimately evaluated against its capacity to improve scientific research.

We end with a personal note. We feel honored for the opportunity to contribute a foreword to this excellent book. Almost ten years ago, we met a young and brilliant student from Italy, Guglielmo, and welcomed him at the IAS-Research Group in San Sebastian (Spain). We remember Guglielmo's stay at San Sebastian, full of enthusiasm for his research and always eager to discuss with us; we remember his intense dedication to research work during the preparation of his thesis—which, by the way, did not prevent him from developing a no less intense work of dissemination of Italian culture in San Sebastian.

All these factors were not negligible in the way they have contributed to a friendly and fruitful relation with us, which ultimately has also contributed to a successful research result. During all the time we worked with Guglielmo and supervised his PhD work, we have been engaged with and challenged by the intense (and often almost interminable, to give credit to his strength and determination!) discussions we had on case studies and their theoretical and philosophical implications. At the end, these discussions were the core ideas that led to a fruitful collaboration and that first materialized in an excellent PhD thesis supported by several papers and, later, gave rise to this book.

We hope that, like us, the reader of this book comes away enriched and full of enthusiasm for exploring new phenomena and ideas.

University of the Basque Country
Leioa, Biscay, Spain

Alvaro Moreno
Leonardo Bich

Preface

My fascination with functional integration began during my early studies in biochemistry and human physiology while I was a master's student in philosophy. Despite being enrolled in the Faculty of Philosophy, the courses I took at the Faculty of Medicine deeply influenced me. I was intrigued by how the physiology of living beings is explained as the integration of functions across various organizational levels—cells, tissues, and organs. This early interest evolved into a focused exploration during my PhD, where I delved into how cells function as integrated wholes. This research highlighted not only the biological underpinnings of functional integration but also its philosophical implications concerning individuality and autonomy.

Despite its central role in defining organisms and biological individuals, the concept of functional integration presents significant challenges. It is often defined too broadly, encompassing any causal interdependence of functions, which makes it difficult to use it to characterize biological organizations as distinct physiological units. Furthermore, the concept lacks a well-developed theoretical framework. These issues underscore the need for a more precise and comprehensive understanding of functional integration, which this book aims to address by bridging the gap between biology and philosophy.

Inspired by Howard Pattee's principle, this book adopts a case study approach, focusing on systems that balance minimal complexity with maximum conceptual interest. The first objective of this book is to elucidate how the *cell*, the fundamental unit of life, functions as a *physiologically integrated system*. I will explore the organizational conditions that allow a cell to exhibit systemic capacities such as metabolism, regulation, signaling, development, reproduction, and sensorimotor functions. The focus will be on a carefully selected yet sufficiently diverse group of *unicellular organisms*—prokaryotes (bacteria and archaea) and unicellular eukaryotes (including protozoa, unicellular algae, and unicellular fungi)—which, despite their simplicity, offer profound insights into the essential processes that define life.

The second—and fundamental—objective is to evaluate the similarities and differences in functional integration between prokaryotes and eukaryotes. This involves examining how functional integration contributes to our understanding of biological *individuality* by determining the type of integration necessary for prokaryotic and eukaryotic unicellular organisms to operate as both *physiological* and *evolutionary units*. Additionally, I will explore how functional integration influences the constitutive and interactive processes, specifically the *autonomy*, of these unicellular organisms.

The third and final purpose is to formulate a theoretical proposal for functional integration in the transition from prokaryotic to eukaryotic cells (*eukaryogenesis*), bridging both biology and philosophy. I aim to provide a qualitative characterization of physiological integration that contributes to a biological theory of cell organization and a philosophical understanding of the ontological status of biological individuals.

The key research questions of this book are framed as follows:

1. How can simple functional structures combine to form more complex functional structures within a cell?
2. Which mechanisms and processes must be integrated to render a cell a physiologically and evolutionarily autonomous unit?
3. What are the similarities and differences between eukaryotic and prokaryotic cells concerning functional integration?

After situating functional integration within current philosophical debates in the philosophy of biology (Chaps. 1-2), Chap. 3 addresses the first question by examining the functional relationship between constitutive, regulatory, and signaling mechanisms. This chapter provides a conceptual foundation for addressing the second and third questions. Chapter 4 presents the main theories of eukaryogenesis and connects evolutionary considerations with the philosophical issues of functional integration and biological autonomy. Chapters 5-8 of the book delve into questions 2 and 3, using eukaryogenesis and the comparison between prokaryotes and eukaryotes as central case studies.

Eukaryogenesis and the comparison between prokaryotes and eukaryotes serve two primary aims: first, to characterize and contrast the types and degrees of functional integration in bacteria, archaea, and unicellular eukaryotes, thereby establishing a theoretical framework for understanding functional integration across these organisms; second, to explore the transition from prokaryotic individuality to the complexity of eukaryotic cells, focusing on key biological processes such as endosymbiosis, endosymbiotic gene transfer, and the invagination of internal membranes. While other cases, like the emergence of functional integration in multicellular organisms or holobionts, present compelling alternatives, practical considerations underscore the choice of eukaryogenesis. Indeed, the shift from prokaryotes to eukaryotes is not only scientifically significant but also exceptionally detailed, enriched by the extensive body of research on eukaryogenesis conducted to date.

It is important to note that this book does not aim to propose new phylogenetic hypotheses about eukaryogenesis. Instead, it analyzes current hypotheses and theories, often using phylogenetic analysis, to characterize functional integration in unicellular organisms and develop plausible hypotheses about key milestones in eukaryogenesis. This is done by comparing current prokaryotic and unicellular eukaryotic organizations. Thus, this book seeks to contribute not only to the ongoing studies of the origin of eukaryotes in evolutionary biology but also to the theoretical characterization of physiological integration in cell biology.

This study examines functional integration in eukaryogenesis through an *organizational framework* that places biological phenomena and properties within their specific contexts. By analyzing both structural features and physicochemical properties, this approach seeks to provide a comprehensive understanding of biological systems. This framework integrates *methodological naturalism* with a *normative evaluation* of biological sciences.

Methodological naturalism in this study involves exploring the ontology of natural phenomena and properties using the analytical and conceptual tools from evolutionary biology, cell biology, and microbiology. The normative aspect evaluates how naturalistic descriptions shape our understanding of functional integration and the broader implications that emerge from these descriptions.

The normative component of the study comprises two main objectives: first, to critically assess the concept of functional integration within the philosophical discourse on biological individuality and autonomy; and second, to develop a comprehensive framework that connects biological and philosophical perspectives in explaining functionally integrated individuals.

The chapters of this book explore the concept of functional integration by testing several key hypotheses. First, functional integration in a cell begins with the interdependence between metabolic and genetic processes, along with signaling and regulatory mechanisms (Chap. 3). This foundational integration is essential for achieving more complex levels of cellular organization. Second, symbiotic associations can lead to various forms of functional integration by imposing specific structural constraints on the behavior of individual components (Chap. 5). Third, the internal partitioning of space through membranes improves the regulation of molecular and metabolic flows within the cell, necessitating a new systemic regulation and physiological integration of intracellular communication (Chap. 6). Fourth, prokaryotic cytoskeletal-like proteins and the eukaryotic cytoskeleton facilitate coordinated sensorimotor capacities, organization of intracellular space, and coordination between developmental and reproductive phases (Chap. 7). Finally, functional integration is crucial for enabling collective reproduction (Chap. 8).

The order of the chapters in this book clearly reflects the five working hypotheses, demonstrating a gradual progression from the simplest to the more complex levels of functional integration in unicellular organisms. This structure illustrates the structural and functional changes that led to a new functionally integrated organization during the transition from prokaryotic to eukaryotic cells.

Chapter 1 introduces the concept of functional integration, situating it within the context of life sciences and philosophical discussions on functions, mechanisms, and individuality. Chapter 2 explores the autopoietic and organizational traditions and the concept of "biological autonomy," highlighting the connections between autonomy, functional integration, and biological teleology. Chapter 3 examines the relationship between metabolic and genetic processes in cells, including intra- and extracellular signals and cellular regulatory mechanisms. Chapter 4 presents the primary hypotheses on eukaryogenesis and addresses the issue of autonomy and functional integration within this evolutionary debate. Chapter 5 investigates the type and degree of functional integration in two fundamental forms of prokaryotic collective organizations: bacterial and archaeal colonies (biofilms) and the endosymbiotic relationship between two bacteria. Chapter 6 explores how the division of the intracellular space by membranes impacts overall cellular functional integration. Chapter 7 examines the role of the cytoskeleton and cytoskeletal-like proteins in achieving functional integration within symbiotic organizations. Chapter 8 assesses the relationship between system-level coordinated reproduction and functional integration. Finally, Chap. 9 provides a summary of the findings from each chapter and discusses the theoretical implications of the book within the current philosophical and biological debate.

In writing this book, my aim is not only to contribute to the theoretical understanding of functional integration but also to provide a framework that bridges biology and philosophy. I hope that this work will inspire others to think critically about the concepts of biological individuality and autonomy and to explore the profound questions that lie at the intersection of these disciplines. Ultimately, my goal is to deepen our understanding of what it means to be a functionally integrated living system and to shed light on the fundamental processes that underpin the complexity of life.

Paris, France Guglielmo Militello

Acknowledgments

This book represents the culmination of a rich and deeply stimulating philosophical journey undertaken during my PhD studies at the University of the Basque Country. I am profoundly grateful to all the members of the IAS Research Group in San Sebastian for their unwavering support and the invaluable feedback they provided throughout the development of this work. My deepest thanks go to Professor Alvaro Moreno and Doctor Leonardo Bich, who not only guided me with exceptional mentorship but also contributed a remarkable foreword to this book. I am also especially grateful to Doctor Kepa Ruiz-Mirazo for his insightful advice on several chapters, particularly those dealing with the theories of eukaryogenesis. I extend my sincere thanks to Professor Juli Peretó for his meticulous reading and constructive suggestions on Chap. 4, as well as to Professor Javier Suarez for his careful review of Chap. 2 and the inspiring conversations that enriched the overall direction of the book. I also wish to acknowledge the support of colleagues from the IHPST in Paris, particularly Doctor Matteo Mossio and Doctor Francesca Merlin, whose valuable insights during my PhD studies significantly influenced some of the content in this book. I am deeply thankful to Florentin and my family for their unwavering love and support during the writing of this book. Finally, I would like to express my gratitude to Springer for their exceptional service, particularly to Christopher Wilby and Solomon George, whose kindness and professionalism were much appreciated. This work was completed without external funding.

Competing Interests The author has no competing interests to declare that are relevant to the content of this manuscript.

Contents

About the Author

Guglielmo Militello earned his PhD in the Philosophy of Biology from the University of the Basque Country (Spain) in 2021. His thesis, which focused on functional integration and individuality in eukaryogenesis, was honored with the university's Best Thesis Award for that year. Following his doctoral studies, he gained experience in medical affairs within the pharmaceutical industry. Currently, he is a postdoctoral researcher at the University of Bordeaux (France) and associated researcher of the *Institut d'Histoire et Philosophie des Sciences et Techniques* in Paris and the University Campus Biomedico in Rome. Militello's research spans three key areas. First, he investigates the theoretical characterization of biological individuality, autonomy, and the emergence of complex life forms during evolutionary transitions. Second, he explores the conceptualization of stem cells and their implications for regenerative medicine and cancer research. Third, his work delves into the potentialities and limitations of anticancer immunotherapies, with a particular focus on the theoretical aspects of tumor immune response and the decisional autonomy of patients undergoing these treatments. His research has been published in several high-impact, peer-reviewed international journals, including *Biology and Philosophy*, *Acta Biotheoretica*, and *Frontiers*.

List of Figures

Chapter 1
Functional Integration in Life Sciences and Philosophy of Biology

Abstract This chapter introduces the book by examining the critical role of functional integration in both the life sciences and philosophical discourse. In life sciences—encompassing cell biology, evolutionary biology, physiology, and pathophysiology—functional integration describes the causal interconnections among an organism's components (such as cells, tissues, and organs) that enable systemic functioning. Philosophically, it is central to debates on biological functions, where it underscores the systemic nature of functions, and in discussions about mechanisms, it highlights the interplay between different mechanisms that drive biological activities. Within the debate on biological individuality, functional integration primarily elucidates the physiological aspects of individuality, although its implications for evolutionary dimensions remain less explored. This chapter argues that the interpretation of functional integration varies across contexts, highlighting its multifaceted significance in understanding the complexity of life and individuality.

Keywords Functional integration · Life sciences · Functions · Mechanisms · Individuality

1.1 Introduction

The concept of functional integration is widely used in the life sciences, playing a key role in the explanations of cell and evolutionary biology, physiology, and pathophysiology. In the philosophy of biology, it is intricately connected to discussions on the nature of biological functions, mechanisms, and individuality. Despite its widespread application, functional integration remains an elusive concept, often encompassing various meanings and entities. This ambiguity, though not diminishing its importance, complicates its practical use.

This chapter addresses four main questions: How is functional integration employed across different domains of life sciences, and what meanings does it take on? How has it been conceptualised within the philosophical debate on biological

G. Militello, *Functional Integration*, History, Philosophy and Theory of the Life Sciences 38, https://doi.org/10.1007/978-3-031-94744-5_1

functions? How is functional integration related to the discourse on mechanisms? How is the concept of biological individuality connected to functional integration?

This introductory chapter addresses these questions by examining how functional integration is employed in life sciences textbooks and foundational works in the history of cell biology. It also reviews key philosophical positions regarding biological functions and mechanisms, exploring the role of functional integration within these frameworks. Additionally, the chapter delves into the multifaceted debate on the nature of biological individuals, highlighting how functional integration influences both the physiological and evolutionary dimensions of individuality.

I argue that functional integration is interpreted differently across various contexts. In life sciences, it generally describes how an organism's components (e.g. cells, tissues, organs) are causally interconnected to support systemic functioning. In discussions of biological functions, functional integration is pivotal for those perspectives that emphasise the systemic aspect of functions. In the debate on mechanisms, it relates more to how mechanisms integrate with one another to facilitate biological activities. Regarding biological individuality, functional integration is typically used to explain the physiological dimension of individuals but is less explored concerning the evolutionary dimension.

The chapter is organised as follows: Sect. 1.2 addresses functional integration in life sciences. Sections 1.3 and 1.4 explore its role in the debates on biological functions and mechanisms, respectively. Section 1.5 analyses the influence of functional integration in discussions on biological individuality. Section 1.6 offers concluding remarks.

1.2 Functional Integration in Life Sciences

In most textbooks on cell biology, physiology, and pathophysiology, functional integration is synonymous with physiological integration. These disciplines intuitively assume that: (a) functional integration is nothing more than the causal interdependence of biological functions; and (b) this causal interdependence explains both the physiology and pathophysiology of organisms.

Such a characterisation makes functional integration an umbrella term encompassing any form of functional coordination and interdependence in any biological system. The concept of functional integration is therefore general and vague, often not examining which physiological dimensions make an organism a functionally integrated whole. Consequently, current definitions of functional integration do not provide criteria to distinguish different biological organisations based on their internal physiological integration.

The concept of functional integration is central to medicine, as *physiology* and *pathophysiology* are grounded in the idea that human beings (and multicellular organisms in general) are physiologically integrated wholes. Medical physiology textbooks explain the physiology of the whole organism in terms of the functional integration within an organ system and among different organ systems[1] (Hall & Hall, 2020). This physiological integration can be lost during a pathological state. For example, heart failure is characterised by the loss of normal physiological integration between heart contraction, cardiac output, and their regulation by the endocrine system and the autonomic nervous system. However, pathological states may lead to a new kind of functional integration in the form of physiological compensations. In heart failure, this results in compensations by the heart, kidneys, endocrine system, and autonomic nervous system (Loscalzo et al., 2022). This new (pathological) integration is what physicians usually call "disease".

In classical molecular biology, the genetic and phenotypic aspects of the cell were explained by studying the circularity between DNA replication, transcription, and translation (Crick, 1958). The central dogma of molecular biology was a reductionist stance that left little room for addressing how a cell is physiologically integrated. In contrast, current approaches in molecular biology, notably in systems biology, study genetic mechanisms as networks that interact with other cellular networks, such as the metabolic and signalling networks (Medina, 2013). The functional integration between molecular functions and cellular systemic conditions entails studying genetic functions in the context of their regulation (i.e. gene regulatory pathways studied by genetics) and their relationship with metabolic pathways studied by biochemistry.

Molecular biology, genetics, and biochemistry are fundamental parts of *cell biology*, which studies cellular processes such as metabolism,[2] intracellular and intercellular communication, and the cell cycle. The latter includes cell growth, DNA replication, transcription and translation, and reproduction. These processes are often defined as integrated for three basic reasons (Alberts, 2022): first, metabolism provides the cell with the energy for all its activities, thus sustaining each phase of the cell cycle; second, both metabolism and the cell cycle hinge on numerous intra- and inter-cellular signals that collectively constitute cellular communication; finally, each cell produces chemical signals that respond to the cell's internal conditions (e.g. metabolic and developmental), so cell-cell communication is related to the metabolic and developmental conditions of each cell.

Despite its frequent use, the concept of functional integration is not conceptualised as such in cell biology and does not explain which cellular processes make a cell a functionally integrated whole. This is especially relevant when investigating the similarities and differences among unicellular organisms: what permits comparison of different biological organisations is not only their morphological and

[1] In neurophysiology, "functional integration" denotes how brain regions collaborate to process information and respond to stimuli. This includes the connections between brain areas and how large groups of neurons react to various stimuli.

[2] Metabolism encompasses ongoing processes such as construction, reconstruction, repair, growth, and reproduction within cells (Barandiaran & Moreno, 2008; Moreno & Ruiz-Mirazo, 1999; Morowitz, 1979, 1992; Ruiz-Mirazo et al., 2004). Even cellular interactions ultimately rely on metabolic organisation.

functional differences but also the way their functions are mutually interdependent.

Another noteworthy aspect is the role of functional integration in *evolutionary biology*. This field appears unrelated to physiological integration since evolutionary biology primarily studies evolutionary transitions through comparative phylogenetics, computational phylogenetics, and comparative anatomy. However, evolutionary transitions can be interpreted as global modifications in the functional integration of a certain biological organisation. Indeed, what evolves are not only genes and phenotypic traits but also the functions performed by these traits and the way they are integrated to perform systemic properties (Margulis, 1970; Buss, 1987; Bonner, 1988; Maynard Smith & Szathmáry, 1995). For example, the transition from prokaryotic to eukaryotic cells involved a global reorganisation of the proto-eukaryotic cell, entailing radical morphological and functional changes, including the appearance of organelles, an increase in cell size, the achievement of new functions, or the transformation of existing ones, leading to a completely new functionally integrated organisation.

1.3 The Nature of Biological Functions

The previous section has demonstrated how life sciences employ the term "functional integration", closely intertwined with the concept of biological function, which holds significant theoretical relevance in both life sciences and philosophy. Early philosophical explorations of function by Nagel (1961), Canfield (1963), and Hempel (1965) focused on the teleological dimension of functional statements. Nagel (1961) and Hempel (1965) argued that functional statements fail as scientific explanations because they do not deduce phenomena from general laws, but rather describe specific behaviours of biological systems in finalistic terms.[3]

These works ignited a lively debate on the nature of biological functions, categorised into four main schools of thought: *aetiological*, *dispositional*, *systemic*, and *organisational* accounts. The aetiological approach explains biological functions in terms of their causal history (Griffiths, 1993; Millikan, 1984, 1989; Neander, 1991; Wright, 1973). Aetiological accounts underscore the role of evolutionary history and natural selection in elucidating biological functions.

Wright (1973), pioneering the aetiological approach, posited that biological functions are intrinsic to biological systems. He argued, "saying that the function of X is Z is saying at least that (i) X is there because it does Z or Doing Z is the reason X is there or That X does Z is why X is there. Where 'because', 'reason', and 'why' have an etiological force" (Wright, 1973, p. 157). Despite natural selection ostensibly explaining why certain organisms possess specific functions, functional

[3]According to Nagel and Hempel, functional statements do not fit the nomological-deductive model proposed by Hempel and Oppenheim (1948). Thus, statements like "the heart beats to pump blood" are not deemed scientific by these authors.

explanations exhibit inherent circularity. Thus, "the function of X is Z means (ii) (a) X is there because it does Z, (b) Z is a consequence (or result) of X's being there" (Wright, 1973, p. 161).

Millikan (1984, 1989) highlighted the evolutionary norm distinguishing proper from accidental functions.[4] Proper functions, like the heart's role in pumping blood, result from capacities developed through evolutionary history and transmitted through reproduction. In contrast, accidental functions, such as the use of a spoon for eating soup, arise from human invention.

Neander (1991) defended an aetiological theory wherein biological functions emerge from natural selection acting on morphological traits. She asserted that a proper function enhances the survival and reproductive fitness of organisms' ancestors, leading to the selection of genes expressing such traits. Griffiths (1993) further differentiated biological functions, selected for enhancing organismal fitness through natural evolution, from artifact functions shaped by human intention.

While the aetiological account looks to the past, explaining functions through their evolutionary history, dispositional approaches adopt a forward-looking perspective. They define functions as means-end relationships that enhance an organism's overall fitness, focusing on their anticipated effects on organisms in the future. Bigelow and Pargetter (1987) view functions as dispositions or propensities of individuals to engage in specific activities. They suggest that these dispositions "can be characterised by reference to possibly nonexistent future events" (Bigelow & Pargetter, 1987, p. 189). According to the authors, defining functions as dispositions contributing to organisms' fitness underscores their strong dependency on the environment. The propensity of a trait to enhance survival hinges on its suitability to the organism's natural habitat. Similarly, Adams (1979) and Boorse (1976, 2002) argue that biological functions causally contribute to systemic capacities, aligning with the fundamental goals of survival and reproduction within the system.

The systemic view of functions,[5] proposed by Cummins (1975), emphasises the current systemic role, particularly the physiological role, played by a function within a biological system. Cummins argues that a biological function "is appealed to explain the biological capacities of the organism containing it, and from these capacities conclusions are drawn concerning the chances of survival for organisms of that type" (Cummins, 1975, p. 751). Attributing a function to a specific trait requires recognising this activity as a disposition to behave in a certain way. In Cummins' framework, a function is always contextualised within a system and can be seen as a causal effect generated by the system components, contributing to the

[4] This applies equally to malfunctions, whose origins are only explainable through their evolutionary history.

[5] Cummins' account is sometimes aligned with dispositional views, treating functions as propensities for higher-level capacities. Unlike traditional dispositional accounts, Cummins emphasises the systemic dimension of biological functions. Thus, I find it more suitable to differentiate Cummins' systemic account from other dispositional perspectives.

development of higher-level capacities.[6] For instance, the function of the heart (i.e. pumping blood) is a direct result of its action, which supports a higher-level capacity like circulation in animals.

The organisational accounts present a fourth perspective on biological functions, synthesising elements from both the aetiological and systemic[7] views (Mossio et al., 2009). Central to the organisational account is the idea that the fundamental nature of biological functions can be elucidated by considering their evolutionary history alongside their current role within specific physico-chemical organisations, such as biological systems (see also Bickhard, 2000; Christensen & Bickhard, 2002; McLaughlin, 2008; Schlosser, 1998). Mossio et al. (2009) highlight that the performance of a particular function hinges on the architecture and organisation of the system, thereby contributing to its self-maintenance. Importantly, the authors assert (Mossio et al., 2009, p. 828) that for a biological function to manifest:

1. The function must contribute to the self-maintenance of the system.
2. The action of each component parts (or trait) of the system must be constrained (or limited) by the action of the other component parts of the system.
3. The system must be internally differentiated.

Condition (1) signifies that a biological function, such as the heart's pumping of blood, is essential for the overall physiology of the organism, enabling its survival. Condition (2) indicates that each biological function is interconnected and reliant on other functions within the organism. Therefore, the pumping of blood is influenced not only by the cardiac muscle's contraction and relaxation but also by other biological traits.[8] Finally, condition (3) stipulates that the system generates various structures that contribute diversely to its maintenance.

The aforementioned perspectives have thoroughly explored the essence of biological functions and their explanations. However, addressing the concept of functional integration within these frameworks proves challenging. Functional integration is either entirely absent from philosophical analysis or, when present, not explicitly conceptualised.

Aetiological and dispositional approaches, in particular, do not grapple with the issue of how biological functions are interconnected within organisms. These theoretical models fail to accommodate functional integration for two primary reasons: Firstly, (1) aetiological accounts do not attribute explanatory value to the current physiological interdependence of functions in the evolutionary history of functions. Secondly, (2) dispositional approaches frame biological functions primarily in

[6] Cummins defines functions as follows: "x functions as a φ in s (or: the function of x in s is to φ) relative to an analytical account A of s's capacity to ψ just in case x is capable of φ-ing in s and A appropriately and adequately accounts for s's capacity to ψ by, in part, appealing to the capacity of x to φ in s" (Cummins, 1975, p. 762).

[7] Mossio et al. (2009, p. 814) equate dispositional and systemic approaches, interpreting functions as causal means-end relations within a system.

[8] Examples include sympathetic and parasympathetic heart stimulation by vagus nerves, or the rate of blood flow from veins into the heart.

terms of means-end relationships and their contribution to overall fitness, portraying functions as isolated units rather than integrated and relational activities within a system.

While systemic and organisational accounts emphasise the theoretical significance of the functional interdependence among biological functions within the systemic context of living organisms, they do not delve into how these functions are specifically integrated within organisms. Cummins underscores the systemic dimension of biological functions, yet he does not explicitly link this dimension to the interdependence of functions that underpins physiology. Similarly, organisational accounts recognise that biological beings rely on a causal circularity of functions, termed "organisational closure", which establishes the basis for physiological functions and enables biological systems to self-maintain. However, the concept of functional integration is often equated with "organisational closure" without a deeper exploration into the biological and organisational aspects that constitute functional integration. As such, while these accounts acknowledge the foundational role of functional interdependence, they do not explicitly explore or define functional integration within the broader context of biological systems.

1.4 Mechanistic Integration

Biological mechanisms, as discussed within the new-mechanistic framework originating in the 1990s, offer a modern philosophical analysis rooted in historical perspectives dating back to the seventeenth century. René Descartes, in works like "Discourse on the Method" ([1637] 1978) and "Treatise on Man" ([1664] 2018), introduced a mechanistic view likening organisms to machines, highlighting how anatomical functions mirror mechanical processes. The neo-mechanistic framework has significantly contributed to understanding how material structures can achieve functional integration within biological structures. By focusing on mechanisms, researchers can elucidate how specific biological functions emerge from the interactions of molecular and cellular components, offering insights that bridge philosophical analyses with empirical investigations in contemporary biology.

Similar but slightly different definitions of biological mechanisms have been provided by scholars like Bechtel, Richardson, Machamer, Darden, and Craver, who have focused extensively on cell and molecular biology (Bechtel & Richardson, 1993; Machamer et al., 2000; Darden, 2008, 2009) as well as neuroscience (Craver, 2007). Here are some nuanced perspectives on what constitutes biological mechanisms according to these scholars:

1. "Mechanisms are entities and activities organized such that they are productive of regular changes from start or set-up to finish or termination conditions" (Machamer et al., 2000).
2. "A mechanism for a behavior is a complex system that produces that behavior by the interaction of a number of parts, where the interaction between parts can be

characterized by direct, invariant, change-relating generalizations" (Glennan, 2002, p. S344).

3. "A mechanism is a structure performing a function in virtue of its component parts, component operations, and their organization. The orchestrated functioning of the mechanism is responsible for one or more phenomena" (Bechtel & Abrahamsen, 2005).

These formulations create an ontological ambiguity regarding mechanisms: Are they entities, activities performed by entities, or both? Machamer et al. (2000) suggest mechanisms encompass both perspectives, while Glennan emphasises they are activities, and Bechtel and Abrahamsen argue they are entities.[9] For instance, when considering the catalytic activity of an enzyme, it remains unclear whether the mechanism comprises solely the catalytic activity or both the activity and the enzyme itself.

Despite differences in these formulations, they converge on defining a biological mechanism with four essential features: parts, organisation, phenomena, and causation (Craver & Tabery, 2023). New-mechanistic approaches are grounded in a mereological ontology that distinguishes component parts (also referred to as "entities" or "structures") from the complex systems they comprise. This distinction allows biologists to identify and attribute specific local activities to each part within a biological system. As highlighted by Bechtel and Richardson (1993), biologists analyse biological phenomena by breaking down systems into their component parts and assigning each part distinct mechanisms. Similarly, Glennan (1996) underscores that component parts exhibit "a kind of robustness and reality" due to their specific properties and activities. Thus, proponents of new-mechanistic accounts agree that living organisms are characterized by discrete parts that can be observed and studied through techniques like microscopy or medical imaging. Each part performs one or more mechanisms, elucidating its functional role within the larger system.

The relationship between component parts performing mechanisms and the entire biological system is crucial in neo-mechanistic theory, characterized by various terms such as "complex system" or "organisation", applicable across diverse biological entities from single cells to multicellular organisms. Some scholars emphasise how the spatial and temporal organisation of these parts within biological systems influences their ability to perform mechanisms. Wimsatt (1997) argues that biological parts exhibit specific arrangements, which are not interchangeable and significantly impact their mechanistic functions.[10] Similarly, Bechtel (2007) observes that spatial factors like size, shape, and location, alongside temporal factors such as order, rate, and duration, define the organisational dynamics of mechanisms.

---

[9] In biomedical sciences, a mechanism refers to activities performed by specific entities, aligning with Machamer and colleagues' account.

[10] Wimsatt (1997) expresses this idea by claiming that mechanistic organisations lack aggregativity.

Interestingly, a mechanism is understood as an organisation of functional parts that, depending on their arrangement, achieve a new *integrated* functional result (Bechtel & Abrahamsen, 2005; Levy, 2014; Militello & Moreno, 2018; Wimsatt, 1986). This interdependence of the component parts within a biological system has been highlighted by Fagan, who introduces the term "jointness" to describe how a mechanism's components work together to produce results that cannot be achieved by the same components when separated (Fagan, 2012).[11] For example, in the case of the enzyme-substrate complex, its mechanism—the catalytic activity—is performed jointly by its components. As Fagan explains, "An enzyme-substrate complex forms in virtue of these components' complementary shapes and biochemical properties, given certain spatiotemporal conditions" (Fagan, 2012, pp. 463–464).

There are two fundamental levels of *mechanistic integration*: firstly, a mechanism is an *integrated process* because it results from the coordinated functioning of its component operations (Bechtel & Abrahamsen, 2005). Secondly, mechanisms are *integrated with each other*, meaning they serve as components within larger mechanisms and are themselves composed of smaller mechanisms, creating distinct levels of mechanistic organisation. In this context, integration refers to the philosophical concepts of "constitution" or "parts-whole relationship" (Craver, 2001).

Regarding the first aspect of mechanistic integration, proponents of new-mechanism emphasise that within the same level of biological organisation (e.g. a cell), a mechanism arises from the integration of various component parts and their operations. For example, according to Bechtel and Abrahamsen, the mechanism of the heart pumping blood results from the coordinated operations of its component parts such as the right and left atria, right and left ventricles, mitral valve, tricuspid valve, pulmonary valve, and aortic valve (Bechtel & Abrahamsen, 2005). This integration of mechanisms encompasses spatial and temporal sequences and involves "an active part that initiates or maintains the operation (and may be changed by it) and at least one passive part that is changed by the operation" (Bechtel & Abrahamsen, 2005, p. 424).

Simultaneously, the integration of mechanisms typically involves the functional interdependence of mechanisms operating at various levels of biological organisation (e.g. cells, tissues, organs, organ systems). According to Craver (2001), mechanisms functioning at lower biological levels (e.g. genes or cells) can be regarded as component parts of larger mechanisms operating at higher levels (e.g. organs or organ systems). For instance, revisiting the example of the heart pumping blood, this process results from numerous mechanisms occurring at the level of cardiac cells (such as depolarisation of cell membranes, opening of voltage-gated calcium channels, and changes in the troponin-tropomyosin protein complex) that are integrated to facilitate the contraction and relaxation of cardiac tissue. Consequently, this cyclic movement within cardiac tissue enables the heart organ to effectively

[11] This concept of interdependence contrasts with the modularity of mechanisms, where parts are semi-independent and separable (Menzies, 2012).

pump blood, facilitating blood circulation throughout the body via the cardiovascular system.

Intra- and *inter-level* mechanistic integration elucidates systemic phenomena in living beings. Mechanisms are thus regarded as the epistemic foundation of physiology and pathophysiology. They not only causally generate biological phenomena, such as the phenomenon of cardiac contraction (systole), but also contribute to their self-maintenance. For instance, systole results from a network of interconnected mechanisms. Moreover, the ability to repeat systole hinges on the cyclic repetition of these mechanisms. Therefore, the regularity of mechanisms and their systemic effects is a crucial feature, as noted by Machamer et al. (2000), who defined mechanisms as "productive of regular changes".

Phenomena are fundamentally linked to mechanisms. As highlighted by Glennan (1996, 2002), the essence of a mechanism is its role in explaining a specific phenomenon. Within the neo-mechanistic framework, mechanistic integration manifests as a causal chain where multiple mechanisms interact to bring about a biological phenomenon. These mechanistic causal chains hold both ontological and epistemological significance, providing explanations for biological phenomena. Liberal perspectives on mechanistic causation, articulated by Glennan (1996), equate it to a cause-and-effect relationship. Similarly, Machamer (2004) and Bogen (2005) view causation as an activity that generates particular outcomes.[12] Other influential perspectives include Salmon's (1984) assertion that mechanistic causation involves the exchange of conserved quantities like mass in space-time,[13] and Woodward's (2001, 2003) argument that mechanistic causation can be discerned through experimental manipulations of biological variables.

Thus, the concept of functional integration is central to the new-mechanistic vocabulary, encompassing both the integration of component operations within a mechanism (*intra-level integration*) and the integration of mechanisms to produce a phenomenon (*inter-level integration*). These dual aspects of mechanistic integration underscore the need for a comprehensive understanding of:

(a) The molecular and biochemical mechanisms underpinning biological phenomena.
(b) How these mechanisms are integrated across different levels of biological organisation.

The new-mechanistic framework raises significant questions that this book will explore. First, if functional integration is understood as a characteristic of individuals, what mechanisms must be integrated to transform a mere aggregation into a cohesive individual entity? Second, if functional integration spans various levels of biological organisation, how do these levels connect and interact with each other?

[12] Machamer and Bogen's definitions have been criticized for being overly broad, failing to distinguish between activities and causal relationships (Psillos, 2004; Woodward, 2002).

[13] This view has been critiqued by authors who argue that conserved quantities are crucial for understanding causation in physics but not in biology (Craver, 2007; Glennan, 2002; Millstein, 2006).

These questions highlight ongoing areas of inquiry within mechanistic biology, aiming to deepen our understanding of how biological systems achieve coherence and functionality across different scales of organisation.

1.5 Individuals as Integrated Wholes

In philosophical and biological literature, the concept of an organism has often been equated with that of a biological individual. Both are described as an "integrated system of interdependent structures and functions" (Lwoff, 1966) or an "integrated whole" (Wolvekamp, 1966). The interdependence of functional parts has been interpreted in various ways. Some authors view it as "functional and structural cohesion" (Collier, 2004, p. 13), essential for maintaining boundaries between the individual and its environment (Godfrey-Smith, 2011, p. 71). Others emphasise that the interdependence of subunits involves "high cooperation and very low conflict" (Queller & Strassmann, 2009, p. 3144), directed towards the "good of the whole" (Okasha, 2011, p. 59).[14] Thus, functional integration has primarily served as an explanatory tool for characterising organism-individuals, rather than being the subject of theoretical investigation itself.

The definitions outlined above align well with paradigmatic cases of organisms, encompassing both unicellular[15] and multicellular[16] forms of life. However, some philosophers (Dupré & O'Malley, 2009; Nicholson, 2014; Pradeu, 2010, 2016) argue that the category of biological individual extends beyond that of organism, suggesting that organisms represent a specific subset within a broader spectrum of individuals. There is ongoing debate as to whether certain collective life forms, such as symbiotic associations or colonies of bacteria and insects, should be classified as biological individuals due to their collective integrated behaviour (Bosch & McFall-Ngai, 2011; Theis et al., 2016; Zilber-Rosenberg & Rosenberg, 2008). However, these entities often lack distinct boundaries, cohesive structures, and system-level reproduction characteristic of paradigmatic organisms, thus exhibiting less pronounced functional integration. Consequently, the varying types and degrees of functional integration among different individuals contribute to their conceptual complexity, surpassing that of organisms.

According to Lidgard and Nyhart (2017), the challenge of defining biological individuals dates back to the nineteenth century, where figures like Huxley (1852) and Spencer (1864) encountered difficulties. Over the past two centuries, scholars have proposed 24 distinct criteria for defining biological individuality, spanning

[14] Similar criteria for the relationship between functional integration and organismality are found in Bock (1989), Sober (1991), Ruiz-Mirazo et al. (2000), Lewontin (2000), Godfrey-Smith (2013).

[15] Unicellular life includes bacteria, archaea, and unicellular eukaryotes.

[16] Multicellular eukaryotic organisms consist of multiple cells typically forming tissues, organs, and organ systems.

from propagation and life cycles to causal integration and fitness maximisation.[17] This diversity has created a complex landscape of concepts and theories. While it is impractical to delve into each criterion here due to space constraints, most of them can be broadly categorized into two main groups: *evolutionary* and *physiological* criteria for individuality (see Pradeu, 2016).[18] Each category offers valuable insights into the relationship between functional integration and individuality, elucidating both the potentials and limitations of current understandings in this field.

1.5.1 Evolutionary Individuals

A fundamental epistemic approach to defining biological individuals and distinguishing them from other biological organisations has been to regard them as evolutionary units: entities that evolve over time under the influence of natural selection, thereby serving as a unit (or level) of selection. Lewontin (1970) was the first to introduce the concept of "*units of selection*", arguing that natural selection operates on the traits of individuals (phenotypic variation); that different phenotypes lead to varying degrees of adaptation to the environment (differential fitness); and that these fitness differences are heritable, passed down from parents to offspring (Lewontin, 1970, p. 1).[19]

A few years later, Dawkins (1976) introduced the concept of *replication* in defining evolutionary individuals. He coined the term "replicator" to describe a molecule that "would act as a template not for an identical copy, but for a kind of 'negative', which would in its turn re-make an exact copy of the original positive" (Dawkins, 2006, p. 16). According to Dawkins, replicators are capable of self-replication and influence their own survival. While DNA molecules are the most well-known examples of replicators, Dawkins suggested that entities broader than genes could also function as replicators, raising questions about which biological entities qualify. Furthermore, Dawkins posited that replicators, rather than biological individuals, are the primary objects of natural selection, reflecting a reductionist perspective that natural selection operates at the level of replicators rather than individuals.

---

[17] Definitional criteria for individuality include propagation, sexual life cycle, demarcated phases of life cycle, causal integration, being a part of a biological hierarchy, exhibiting adaptation, germ/soma distinction, genetic homogeneity, immunological integration, autonomy, spatio-temporal continuity, spatio-temporal boundedness, indivisibility, bottleneck in life cycle, unit of selection, fitness maximisation, high cooperation among the parts, genetic/epigenetic regulatory interaction, persistence, mechanisms of diminishing cell lineage conflict, non-genetic microbial interaction, heritable variation of fitness, cognition, and high interaction among groups of living entities (Lidgard & Nyhart, 2017, pp. 19–21).

[18] Pradeu's distinction clarifies differences between accounts of biological individuality, including ecological individuals discussed in Chap. 5.

[19] Lewontin defines a unit of selection as "any entities in nature that have variation, reproduction, and heritability" (1970, p. 1).

Dawkins' concept of replicators was expanded upon by Hull (1980), who emphasised that replicators create copies through interactions with their environment. Hull distinguished two types of entities subject to natural selection: *replicators*, which transmit their structure directly through replication, and *interactors*, which interact as cohesive wholes with their environment in a manner that affects differential replication (Hull, 1980, p. 318). Natural selection, Hull argued, involves the differential perpetuation of replicators due to the differential extinction and proliferation of interactors they produce (Hull, 1980, p. 318). Hull's characterisation of interactors highlights two key aspects: firstly, various biological entities—genes, cells, and organisms—operate as different levels of interaction[20] (Hull, 1980, p. 325); secondly, interaction with the environment necessitates being a cohesive whole, implying a requisite degree of functional integration. Therefore, Hull suggests that evolutionary individuality hinges on physiological individuality and functional integration, where entities must exhibit integrated physiological capacities to interact effectively with their environment (Hull, 1980).

Griffiths and Gray (1994) argued that a unit of replication involves developmental systems and processes, describing the developmental process or *life cycle*[21] as a series of events that repeat within lineages. They defined the developmental system as the structured set of resources used to reconstruct the life cycle in each generation (Griffiths & Gray, 1994, p. 304). The authors proposed that evolutionary individuality requires an expanded concept of inheritance, emphasising how past generations shape the developmental context of their successors (Griffiths & Gray, 1994, p. 283). Griffiths and Gray highlighted the cyclic nature of developmental processes, where life cycles enable the reconstruction of functional lineage structures from relatively simple resources (Griffiths & Gray, 1994, p. 296).

Similarly, Griesemer (2000, 2016) emphasised the connection between *development* and *reproduction*, arguing that a biological individual is defined by its ability to reproduce through developmental processes, generating offspring that proliferate via reproduction (Griesemer, 2000, p. S362). Central to Griesemer's perspective is the concept of "material overlap", where reproduction involves "bonds of material continuity, not merely resemblance or formal information transmission" (Griesemer, 2016, p. 807). This close integration of development and reproduction has prompted philosophers and biologists (Rainey & Kerr, 2010; Sousa, 2005; Wilson, 2004) to highlight the significance of life cycles in understanding the nature of biological individuals.

The significance of reproduction has also been highlighted in Godfrey-Smith's account (Godfrey-Smith, 2009, 2011, 2013) of *Darwinian individuals*. These individuals exhibit trait variation that is heritable and confers a reproductive advantage, making them active reproducers of heritable traits and members of Darwinian populations (Godfrey-Smith, 2013, pp. 19–20). Godfrey-Smith distinguishes three types

[20] Hull (1980) suggests entities like genes or cells can act as both replicators and interactors.

[21] In scientific literature, "life cycle" refers to processes of development and reproduction forming a cycle.

of individuals based on their mode of reproduction: first, those reproducing as part of a larger unit (scaffolded reproducers); second, those reproducing independently (simple reproducers); and third, those with specialised systems for reproduction (collective reproducers) (Godfrey-Smith, 2009, pp. 87–89). Collective reproducers are defined by reproductive specialisation, bottleneck stages, and integration. In this context, Godfrey-Smith underscores the importance of functional integration in defining evolutionary individuals, although he applies this concept mainly to collective reproducers and not to other biological entities. Additionally, the notion of functional integration remains rather general, often equated with the idea of a "cohesive whole".

The problem of functional integration in evolutionary individuals has been clearly addressed by Shapiro (2011), who argues that the emergence of new morpho-functional traits ("evolutionary novelties") is, on the one hand, the result of genetic mutations[22] and, on the other, a consequence of new selective pressures exerted by the functional integration of pre-existing traits. This integration "places constraints on the kinds of additional inventions that will prove adaptive" (p. 144). As a result, functional integration is portrayed as a dynamic state that is both a cause and an effect of evolutionary individuals: it is a cause because it selects new traits, thereby enabling Lewontin's phenotypic variation; it is an effect because it results from a process of adaptation (Lewontin's differential fitness).

Similarly, Lala et al. (2024) argue that functional integration is a characteristic of evolutionary organisms that promotes phenotypic variation and is shaped by adaptation. Compared to Shapiro (2011), Lala and colleagues emphasise that functional integration is closely linked to developmental processes, which "enable organisms to maintain functional integration even under sometimes severe perturbation" (pp. 98–99). Development is not merely a sequence of morpho-functional changes occurring throughout an organism's life; rather, it is a key biological feature that connects the genotype space with the phenotype space "by determining phenotypic responses to genetic and environmental perturbations, as well as which traits are expressed together and what impact they have on reproductive success or survival" (p. 228).

It is worth emphasising that in both Shapiro's and Lala and colleagues' accounts, functional integration is treated as an *explanans* of evolutionary individuals rather than being explicitly defined: it serves as a concept that helps explain what evolutionary individuals are and how they adapt to their environment. Accordingly, they adopt the basic meaning of functional integration as the functional interdependence among morphological traits, without considering it as something that itself requires explanation (i.e. the *explanandum*).

Santelices (1999) proposed a gradualist view of biological individuality, characterised by *genetic uniqueness*, *homogeneity*, and *autonomy*. Genetic uniqueness

[22] Shapiro's view of genetic mutations includes both multiple DNA small random mutations (e.g. localised point mutations and horizontal gene transfer) and "natural genetic engineering", which designates "the built-in processes of self-modification that have operated over the course of evolution" (p. 132).

refers to the presence or absence of a unique genome; homogeneity refers to the number of genetic changes during ontogeny;[23] and autonomy refers to the capacity to act independently "with respect to the forces of nature [...] in a way that allows it to reproduce" (Santelices, 1999, p. 152). The co-presence of these attributes "characterises each kind of individual as a unit of selection and by its relationship with the environment" (Santelices, 1999, p. 154). However, these attributes can vary along a spectrum, leading to different possible combinations. Unitary individuals (e.g. animals) exemplify a type of individuality that meets all three traditional criteria.

Santelices' account clarifies fundamental aspects of individuality and highlights the existence of different degrees of individuality. Nevertheless, his view has limitations: (a) it is restricted to multicellular organisms and does not encompass unicellular ones; and (b) it risks offering an overly liberal view of individuality, lacking clear boundaries between different types of biological organisations. Furthermore, in Santelices' framework, evolutionary individuality is reduced to genetic uniqueness/homogeneity and autonomous interaction with the environment. However, these capacities necessitate a form of cohesive unity, achieved through the integration of functional parts.

We can now draw some conclusions about the functional integration of evolutionary individuality. First, although the notion of replicator does not entail functional integration, that of interactor requires a certain degree of cohesiveness, allowing it to interact with the environment in a coordinated way. Secondly, the importance of the development-reproduction axis (or life cycles) raises two intriguing questions about the functional integration of an individual: how are development and reproduction functionally interdependent in terms of biological mechanisms? And how do development and reproduction depend on the global physiological capacities (e.g. metabolism) of an individual, and what functional contributions do they make to an individual's physiology? Finally, some biologists (Doolittle, 2013; Martins & Locke, 2015) have noted that the absence of genetic uniqueness and homogeneity can enhance coordinated behaviour and collective functions, as seen in bacterial colonies. These colonies share many genes through lateral gene transfer[24] and often exhibit highly coordinated behaviours, such as in biofilms. These theoretical accounts suggest that the functional integration of an individual also depends on a specific genetic configuration, which in turn hinges on a particular biological organisation that can constrain gene behaviour and genetic expression.

[23] Some individuals (e.g. some plants, algae) may lack a unique genome due to clonal propagation (Santelices, 1999, p. 152), while others (e.g. tumour cells) may exhibit genetic uniqueness but lack genetic homogeneity due to marked genotype changes during development (Santelices, 1999, p. 153).

[24] Lateral (or horizontal) gene transfer refers to genetic material transfer between organisms, distinct from vertical transfer via reproduction.

1.5.2 Physiological Individuals

The second epistemic strategy to define biological individuals is to consider them as functional or physiological units. A basic criterion for physiological individuality is *metabolism*, which involves the biochemical transformation of macromolecules. Metabolic processes, organised into metabolic pathways, are divided into catabolic and anabolic processes. Catabolic processes produce energy by breaking down macromolecules, while anabolic processes consume energy by building up macromolecules. Metabolism plays a pivotal role in the self-maintenance of organisms, keeping them far from thermodynamic equilibrium.[25]

Jonas (1966) considered metabolism as the basis for all biological functions, describing it as a dual process: it enables the organism to be intimately connected with its environment through the transformation of nutritional elements while also establishing a boundary between the interior and exterior of the organism. Consequently, being a metabolic unit is seen as a distinguishing feature of individual organisms. In contrast, Dupré and O'Malley have emphasised that metabolism is "typically a collaborative activity" (Dupré & O'Malley, 2009, p. 13), involving different organisms in symbiotic relationships. They suggest that metabolic interactions among various organisms allow for the possibility of defining biological individuals not by organismic features but by metabolic relationships. Taking this further, Landecker (2017) highlighted that metabolism reveals the intimate interconnection among organisms, from unicellular to multicellular forms of life. Therefore, defining an individual as a metabolic unit implies a dependency on other organisms.

In addition to metabolism, the *immune system* plays a fundamental role in constituting the individual as a functional unit, as immune interactions "are systemic (as opposed to local) and [...] responsible for the acceptance or rejection of constituents in the organism" (Pradeu, 2010, p. 258). While metabolic criteria are quite liberal and risk overlooking the distinctions between different types of physiological individuals, "the immune system constitutes a discrimination mechanism, accepting some entities in the organism and rejecting others, thus participating decisively in the delineation of the organism's boundaries" (Pradeu, 2016, p. 803). Moreover, the immune system provides three main senses of individuality: uniqueness, delineation (i.e. its boundaries), and persistence (Pradeu, 2016, p. 803). These criteria offer a strong view of immunological individuality, defining "an immunological individual as a functionally integrated whole made up of heterogeneous constituents that are locally interconnected by strong biochemical interactions and controlled by systemic immune interactions" (Pradeu, 2016, p. 804). Furthermore,

[25] As discussed in Chap. 2, living organisms adhere to thermodynamic laws. The second law states entropy (disorder) increases in isolated systems. Living beings are open systems exchanging energy and matter, remaining far from thermodynamic equilibrium via catabolic and anabolic processes. At thermodynamic equilibrium, no energy or matter exchanges occur, rendering the biological system non-functional.

Pradeu's account suggests that symbiotic associations (e.g. a host and its microbes) can be considered a unique immunological individual because microbes interact regularly with the immune system and are not eliminated by it. Thus, they are "part of the physiological individual" (Pradeu, 2016, p. 805).[26]

Another important criterion for defining physiological individuality is considering it as a functionally integrated organisation, where the relationship between *parts* and the *whole* is crucial. Some authors (Korn, 2002; Zylstra, 1992) have emphasised that individuals exhibit a hierarchy of entities, ranging from atoms to species, which determine mutually related levels of organisation. Each level of organisation contributes to the constitution of higher levels, and higher levels, in turn, constrain the functions of lower levels. A fundamental aspect of individuals is that they constrain the behaviour of their component parts while being constrained by the components of their ecological niche (Korn, 2002). According to Korn, an individual is "an entity that hierarchically is both made by one kind of process and is a building block of something larger made by a different type of process" (Korn, 2002, p. 203). Similarly, Zylstra (1992) argued that individuals are enkaptic structural wholes, consisting of "a relation between different types of individuality structures (wholes)" (Zylstra, 1992, p. 126). For example, a cell is an enkaptic structural whole because it results from the relationships between the molecules composing it, "exhibiting an internal teleonomy independent of that of the cell" (Zylstra, 1992, p. 127). In other words, Zylstra suggests that individual organisms resemble matryoshkas, where there is not merely a hierarchy of part-whole relationships but rather a hierarchy of enkaptic structural wholes.

In all individuals where the component parts have relative autonomy from one another, a major challenge the overall system must overcome to achieve functional integration is the potential conflict among these parts. This can be observed in symbiotic associations as well as in the origin of multicellularity, where single cells may compete with one another from a metabolic or immunological perspective. This is why Michod and Roze (1997) argued that the evolution of multicellular organisms—and, more generally, of individuals—requires an increase in cooperation among cells and "the emergence of higher level functions that restrict the opportunity for conflict within and the ensure the continued cooperation of lower-level units" (p. 857). In a similar vein, Queller and Strassmann (2009) claimed that the parts of an individual must exhibit a *high* degree of *cooperation* and a *low level of conflict* to form a functionally integrated organisation.[27] Notably, Queller and Strassmann provide a social definition of biological individuality, aiming to theoretically underpin collective forms of life. However, their argument also applies to

[26] In a similar vein, Gilbert and Tauber (2016) argue that immunological individuality supports an ecological view, considering individuals as combinations of several agents.

[27] The authors clarify that while conflict may hinder cooperation, they are not mutually exclusive (Queller & Strassmann, 2009, p. 3151).

explaining how an individual organism functions as a cohesive whole.[28] Moreover, their criterion can support the concept of evolutionary individuality, as "control of conflict is viewed as one of the major prerequisites for a major [evolutionary] transition" (Queller & Strassmann, 2009, p. 3151).

Although high cooperation and low conflict are crucial for defining both physiological and collective individuals, Folse and Roughgarden (2010) highlight that *functional differentiation* is essential for the viability of complex multicellular systems and colonies of social insects.[29] They propose that an individual organism is characterised by three main features: alignment of fitness, export of fitness through germ-soma specialisation, and adaptive functional organisation. The alignment of fitness implies that the component parts of a biological system aim to enhance the fitness of the whole. The export of fitness to the organism level signifies that the component parts relinquish their individual fitness in favour of the organism's fitness through functional specialisation. Lastly, adaptive functional organisation refers to the division of labour and coordinated action within the organism.

Arnellos et al. (2014) argue that a certain level of *internal functional differentiation* enables a system to adapt to external constraints and replicate this structure through generations. This functional differentiation requires a specific type of functional integration, identified with a regulatory control subsystem that "(1) is capable of modulation of intracellular epigenetic mechanisms; and (2) can do so in a recurrent way: i.e., they can trigger off generation, within the system, of ever new similar constraints" (p. 857). The role of epigenetic regulatory pathways in achieving functional integration is evident even in the simplest organisms such as prokaryotes,[30] although their capacity to induce differentiation is notably weaker compared to that of eukaryotic cells (Arnellos et al., 2014, p. 876).

The accounts of physiological individuality discussed above offer valuable insights into functional integration. They highlight several critical dimensions: metabolic and immunological functions, the interdependence and cooperation among parts, functional differentiation, and epigenetic regulation. These dimensions collectively contribute to the stability and resilience of the physiological system, enabling it to withstand both internal and external challenges. For instance, the metabolic dimension relies on the immune system to combat pathogens, while the immune system depends on metabolic processes to supply energy for immune responses. Both systems are intricately regulated by genetic and epigenetic mechanisms and exhibit functional differentiation. However, a common limitation across these accounts is their tendency to focus on individual aspects of physiological unity rather than considering them holistically. This raises a fundamental question:

[28] Queller and Strassmann's view aligns with Gardner and Grafen (2009), emphasizing parts work cooperatively in organisms (p. 3144).

[29] Folse and Roughgarden (2010) critique physiological and genetic individuality criteria in favour of evolutionary criteria, applicable to physiological individuals.

[30] Although "prokaryote" is now replaced by bacteria and archaea, I continue using the term for simplicity, referring to unicellular organisms in both domains.

how do these various dimensions interact and coordinate to produce a physiologically integrated whole?

Lastly, the majority of the accounts discussed above distinguish physiological unity from evolutionary individuality, thereby categorising many collective life forms as complete individuals. However, separating the physiological dimension from the evolutionary one poses challenges when explaining the emergence of new forms of individuality during natural evolution. If a biological organisation, classified as a physiological individual, cannot persist across generations, how do we account for the emergence of new individuals from it? This necessitates further distinctions between categories of individuals with evolutionary potential and those lacking it.

1.6 Concluding Remarks

To conclude, the concept of functional integration plays a crucial role in both life sciences and philosophy. In both domains, it has often been used intuitively or implicitly rather than explicitly conceptualised. In life sciences, functional integration refers to the causal interdependence of biological mechanisms operating at various levels (e.g. genes, cells, tissues, organs), giving rise to systemic capacities such as cellular physiology or the coordinated functions of organ systems in multicellular organisms. I have demonstrated that pathophysiological conditions are frequently understood in terms of functional integration, where diseases result from dysregulated integrated mechanisms. In evolutionary biology, functional integration is equally significant, as the emergence of new forms of individuality during evolution can be viewed as the emergence of novel types of functionally integrated organisations.

The nature of biological functions, mechanisms, and individuals has been extensively debated in the philosophy of biology. Functional integration plays a pivotal role in each of these discussions. In the debate concerning the nature of biological functions, functional integration receives less attention in aetiological approaches but is prominently featured in systemic accounts. Systemic perspectives emphasise that biological functions derive their nature from their current roles within a biological system and their interdependence with other functions. I have highlighted that an organisational approach to functions as proposed by Mossio et al. (2009), which aims to synthesise aetiological and systemic perspectives, offers a robust framework for theorising about biological functions.

In the neo-mechanistic debate, the concept of functional integration is often assimilated into that of mechanistic integration. I have underscored that mechanistic integration manifests in two distinct senses: firstly, as the intra-level integration of activities to form a mechanism; and secondly, as the inter-level integration of mechanisms to generate systemic phenomena. Biological mechanisms adhere to a spatiotemporal order and are inherently embedded within specific biological organisations. Therefore, comprehending how mechanisms are functionally integrated to produce systemic phenomena necessitates a deep understanding of biological organisation. This includes aspects such as architectural structure, inter-level relationships, and regulatory processes. In essence, elucidating biological organisation is crucial for explaining how mechanisms integrate to manifest systemic phenomena.

Functional integration is intricately intertwined with the concepts of organismality and individuality, forming a dialectic relationship between them. Organismality generally denotes a robustly integrated system of interdependent parts and functions, while individuality can be understood either as an evolutionary unit (evolutionary individual) or a physiological unit (physiological individual). I have posited that evolutionary individuals necessitate a certain degree of functional integration. For instance, the developmental-reproduction axis, crucial for a system to reproduce and pass on its traits to offspring, is intricately integrated with other biological dimensions such as metabolism. Additionally, physiological individuals emerge from the integration of several dimensions—metabolic, immunologic, genetic, and epigenetic—enabling the system to maintain stability and withstand both external and internal disruptions effectively.

Functional integration emerges as a fundamental aspect of biological individuals, comprising interconnected systemic functions reliant on integrated molecular and biochemical mechanisms. This underscores why functional integration features prominently in explanations within life sciences. However, functional integration typically serves as an explanatory factor rather than the object of inquiry itself, lacking a robust theoretical foundation. Several pivotal questions thus arise: Which systemic functions and mechanisms must integrate to constitute a physiological individual? How does physiological individuality relate to evolutionary individuality? What forms of functional integration are necessary to transition a physiological individual into an evolutionary one? How can understanding functional integration illuminate the emergence of new forms of individuality in natural evolution? These pressing questions constitute the primary research focus of the forthcoming chapters of this book.

References

Adams, F. R. (1979). A goal-state theory of function attributions. *Canadian Journal of Philosophy, 9*(3), 493–518. https://doi.org/10.1080/00455091.1979.10716265

Alberts, B. (2022). *Molecular biology of the cell* (7th ed.). Norton & Company.

Arnellos, A., Moreno, A., & Ruiz-Mirazo, K. (2014). Organizational requirements for multicellular autonomy: Insights from a comparative case study. *Biology & Philosophy, 29*(6), 851–884. https://doi.org/10.1007/s10539-013-9387-x

Barandiaran, X., & Moreno, A. (2008). Adaptivity: From metabolism to behavior. *Adaptive Behavior, 16*(5), 325–344. https://doi.org/10.1177/1059712308093868

Bechtel, W. (2007). Biological mechanisms: Organized to maintain autonomy. In F. C. Boogerd, F. J. Bruggeman, J.-H. S. Hofmeyr, & H. V. Westerhoff (Eds.), *Systems biology* (pp. 269–302). Elsevier. https://doi.org/10.1016/B978-044452085-2/50014-0

Bechtel, W., & Abrahamsen, A. (2005). Explanation: A mechanist alternative. *Studies in History and Philosophy of Science Part C: Studies in History and Philosophy of Biological and Biomedical Sciences, 36*(2), 421–441. https://doi.org/10.1016/j.shpsc.2005.03.010

Bechtel, W., & Richardson, R. C. (1993). *Discovering complexity: Decomposition and localization as strategies in scientific research* (1st ed.). MIT Press.

Bickhard, M. H. (2000). Autonomy, function, and representation. *Communication and Cognition-Artificial Intelligence, 17*(3–4), 111–131.

Bigelow, J., & Pargetter, R. (1987). Functions. *Journal of Philosophy, 84*(4), 181–196. https://doi.org/10.2307/2027157

Bock, W. J. (1989). Organisms as functional machines: A connectivity explanation. *American Zoologist, 29*(3), 1119–1132. https://doi.org/10.1093/icb/29.3.1119

Bogen, J. (2005). Regularities and causality; generalizations and causal explanations. *Studies in History and Philosophy of Science Part C: Studies in History and Philosophy of Biological and Biomedical Sciences, 36*(2), 397–420. https://doi.org/10.1016/j.shpsc.2005.03.009

Bonner, J. T. (1988). *The evolution of complexity by means of natural selection*. Princeton University Press.

Boorse, C. (1976). What a theory of mental health should be. *Journal for the Theory of Social Behaviour, 6*(1), 61–84. https://doi.org/10.1111/j.1468-5914.1976.tb00359.x

Boorse, C. (2002). A rebuttal on functions. In A. Ariew, R. C. Cummins, & M. Perlman (Eds.), *Functions: New essays in the philosophy of psychology and biology* (pp. 63–112). Oxford University Press.

Bosch, T. C. G., & McFall-Ngai, M. J. (2011). Metaorganisms as the new frontier. *Zoology, 114*(4), 185–190. https://doi.org/10.1016/j.zool.2011.04.001

Buss, L. W. (1987). *The evolution of individuality*. Princeton University Press.

Canfield, J. (1963). Teleological explanation in biology. *British Journal for the Philosophy of Science, 14*(56), 285–295. https://doi.org/10.1093/bjps/xiv.56.285

Christensen, W., & Bickhard, M. H. (2002). The process dynamics of normative function. *The Monist, 85*(1), 3–28. https://doi.org/10.5840/monist20028516

Collier, J. D. (2004). Self-organization, individuation and identity. *Revue Internationale de Philosophie, 228*(2), 151–172. https://doi.org/10.3917/rip.228.0151

Craver, C. F. (2001). Role functions, mechanisms, and hierarchy. *Philosophy of Science, 68*(1), 53–74. https://doi.org/10.1086/392866

Craver, C. F. (2007). *Explaining the brain: Mechanisms and the mosaic unity of neuroscience*. Oxford University Press.

Craver, C. F., & Tabery, J. (2023). Mechanisms in science. In E. N. Zalta & U. Nodelman (Eds.), *The Stanford encyclopedia of philosophy*. https://plato.stanford.edu/archives/fall2023/entries/science-mechanisms/

Crick, F. H. (1958). On protein synthesis. *Symposia of the Society for Experimental Biology, 12*, 138–163.

Cummins, R. E. (1975). Functional analysis. *Journal of Philosophy, 72*(November), 741–764. https://doi.org/10.2307/2024640

Darden, L. (2008). Thinking again about biological mechanisms. *Philosophy of Science, 75*(5), 958–969. https://doi.org/10.1086/594538

Darden, L. (2009). Discovering mechanisms in molecular biology. In J. Meheus & T. Nickles (Eds.), *Models of discovery and creativity* (pp. 43–55). Springer. https://doi.org/10.1007/978-90-481-3421-2_3

Dawkins, R. (1976). *The selfish gene*. Oxford University Press.

Dawkins, R. (2006). *The selfish gene: 30th anniversary edition*. Oxford University Press.

Descartes, René. ([1637] 1978). *Discours de la méthode*. Flammarion.

Descartes, René. ([1664] 2018). *L'Homme*. Flammarion.

Doolittle, W. F. (2013). Microbial neopleomorphism. *Biology and Philosophy, 28*(2), 351–378. https://doi.org/10.1007/s10539-012-9358-7

Dupré, J., & O'Malley, M. (2009). Varieties of living things: Life at the intersection of lineage and metabolism. *Philosophy, Theory, and Practice in Biology, 1*(3), e003. https://doi.org/10.3998/ptb.6959004.0001.003
Fagan, M. B. (2012). The joint account of mechanistic explanation. *Philosophy of Science, 79*(4), 448–472. https://doi.org/10.1086/668006
Folse, H. J., & Roughgarden, J. (2010). What is an individual organism? A multilevel selection perspective. *The Quarterly Review of Biology, 85*(4), 447–472. https://doi.org/10.1086/656905
Gardner, A., & Grafen, A. (2009). Capturing the superorganism: A formal theory of group adaptation. *Journal of Evolutionary Biology, 22*(4), 659–671. https://doi.org/10.1111/j.1420-9101.2008.01681.x
Gilbert, S. F., & Tauber, A. I. (2016). Rethinking individuality: The dialectics of the holobiont. *Biology and Philosophy, 31*(6), 839–853. https://doi.org/10.1007/s10539-016-9541-3
Glennan, S. (1996). Mechanisms and the nature of causation. *Erkenntnis, 44*(1), 49–71. https://doi.org/10.1007/bf00172853
Glennan, S. (2002). Rethinking mechanistic explanation. *Philosophy of Science, 69*(S3), 342–353. https://doi.org/10.1086/341857
Godfrey-Smith, P. (2009). *Darwinian populations and natural selection*. Oxford University Press.
Godfrey-Smith, P. (2011). Darwinian populations and transitions in individuality. In B. Calcott & K. Sterelny (Eds.), *The evolutionary transitions revisited*. MIT Press.
Godfrey-Smith, P. (2013). Darwinian individuals. In F. Bouchard & P. Huneman (Eds.), *From groups to individuals: Evolution and emerging individuality* (pp. 17–36). MIT Press.
Griesemer, J. (2000). Development, culture, and the units of inheritance. *Philosophy of Science, 67*(S3), S348–S368. https://doi.org/10.1086/392831
Griesemer, J. (2016). Reproduction in complex life cycles: Toward a developmental reaction norms perspective. *Philosophy of Science, 83*(5), 803–815. https://doi.org/10.1086/687865
Griffiths, P. E. (1993). Functional analysis and proper functions. *British Journal for the Philosophy of Science, 44*(3), 409–422. https://doi.org/10.1093/bjps/44.3.409
Griffiths, P. E., & Gray, R. D. (1994). Developmental systems and evolutionary explanation. *Journal of Philosophy, 91*(6), 277–304. https://doi.org/10.2307/2940982
Hall, J. E., & Hall, M. E. (2020). *Guyton and hall textbook of medical physiology* (14th ed.). Elsevier.
Hempel, C. G. (1965). *Aspects of scientific explanation and other essays in the philosophy of science*. The Free Press.
Hempel, C. G., & Oppenheim, P. (1948). Studies in the logic of explanation. *Philosophy of Science, 15*(2), 135–175.
Hull, D. L. (1980). Individuality and selection. *Annual Review of Ecology, Evolution, and Systematics, 11*, 311–332. https://doi.org/10.1146/annurev.es.11.110180.001523
Huxley, T. H. (1852). Upon animal individuality. In *Proceedings of the Royal Institution of London (1851–1854)* (pp. 184–189).
Jonas, H. (1966). *The phenomenon of life: Toward a philosophical biology*. Harper & Row.
Korn, R. W. (2002). Biological hierarchies, their birth, death and evolution by natural selection. *Biology and Philosophy, 17*(2), 199–221. https://doi.org/10.1023/A:1015271707591
Lala, K., Uller, T., Feiner, N., Feldman, M. W., & Gilbert, S. F. (2024). *Evolution evolving: The developmental origins of adaptation and biodiversity*. Princeton University Press.
Landecker, H. (2017). 9. Metabolism, autonomy, and individuality. In S. Lidgard & L. K. Nyhart (Eds.), *Biological individuality: Integrating scientific, philosophical, and historical perspectives* (pp. 225–248). University of Chicago Press. https://doi.org/10.7208/9780226446592-010
Levy, A. (2014). Machine-likeness and explanation by decomposition. *Philosophers' Imprint, 14*(6) https://quod.lib.umich.edu/cgi/p/pod/dod-idx/machine-likeness-and-explanation.pdf?c=phimp;idno=3521354.0014.006;format=pdf
Lewontin, R. C. (1970). The units of selection. *Annual Review of Ecology, Evolution, and Systematics, 1*, 1–18. https://doi.org/10.1146/annurev.es.01.110170.000245

Lewontin, R. C. (2000). *The triple helix: Gene, organism, and environment*. Harvard University Press.
Lidgard, S., & Nyhart, L. K. (2017). The work of biological individuality: Concepts and contexts. In S. Lidgard & L. K. Nyhart (Eds.), *Biological individuality: Integrating scientific, philosophical, and historical perspectives* (pp. 17–62). University of Chicago Press.
Loscalzo, J., Fauci, A. S., Kasper, D. L., Hauser, S., Longo, D., & Larry Jameson, J. (2022). *Harrison's principles of internal medicine* (21st ed.). McGraw Hill.
Lwoff, A. (1966). Interaction among virus, cell, and organism. *Science, 152*(3726), 1216–1220. https://doi.org/10.1126/science.152.3726.1216
Machamer, P. (2004). Activities and causation: The metaphysics and epistemology of mechanisms. *International Studies in the Philosophy of Science, 18*(1), 27–39. https://doi.org/10.1080/02698590412331289242
Machamer, P., Darden, L., & Craver, C. F. (2000). Thinking about mechanisms. *Philosophy of Science, 67*(1), 1–25. https://doi.org/10.1086/392759
Margulis, L. (1970). *Origin of eukaryotic cells: Evidence and research implications for a theory of the origin and evolution of microbial, plant, and animal cells on the precambrian earth* (1st ed.). Yale University Press.
Martins, B. M. C., & Locke, J. C. W. (2015). Microbial individuality: How single-cell heterogeneity enables population level strategies. *Current Opinion in Microbiology, 24*(April), 104–112. https://doi.org/10.1016/j.mib.2015.01.003
McLaughlin, P. (2008). *What functions explain: Functional explanation and self-reproducing systems*. Cambridge University Press.
Medina, M. Á. (2013). Systems biology for molecular life sciences and its impact in biomedicine. *Cellular and Molecular Life Sciences, 70*(6), 1035–1053. https://doi.org/10.1007/s00018-012-1109-z
Menzies, P. (2012). The causal structure of mechanisms. *Studies in History and Philosophy of Science Part C: Studies in History and Philosophy of Biological and Biomedical Sciences, 43*(4), 796–805. https://doi.org/10.1016/j.shpsc.2012.05.008
Michod, R. E., & Roze, D. (1997). Transitions in individuality. *Proceedings of the Royal Society B: Biological Sciences, 264*(1383), 853–857.
Militello, G., & Moreno, Á. (2018). Structural and organisational conditions for being a machine. *Biology and Philosophy, 33*(5–6), 35. https://doi.org/10.1007/s10539-018-9645-z
Millikan, R. G. (1984). *Language, thought, and other biological categories: New foundations for realism*. MIT Press.
Millikan, R. G. (1989). In defense of proper functions. *Philosophy of Science, 56*(2), 288–302. https://doi.org/10.1086/289488
Millstein, R. L. (2006). Natural selection as a population-level causal process. *British Journal for the Philosophy of Science, 57*(4), 627–653. https://doi.org/10.1093/bjps/axl025
Moreno, A., & Ruiz-Mirazo, K. (1999). Metabolism and the problem of its universalization. *Biosystems, 49*(1), 45–61. https://doi.org/10.1016/S0303-2647(98)00034-3
Morowitz, H. J. (1979). *Energy flow in biology*. Ox Bow Press.
Morowitz, H. J. (1992). *Beginning of cellular life: Metabolism recapitulates biogenesis*. Yale University Press.
Mossio, M., Saborido, C., & Moreno, A. (2009). An organizational account of biological functions. *British Journal for the Philosophy of Science, 60*(4), 813–841. https://doi.org/10.1093/bjps/axp036
Nagel, E. (1961). *The structure of science: Problems in the logic of scientific explanation*. Harcourt, Brace & World.
Neander, K. (1991). Functions as selected effects: The conceptual analyst's defense. *Philosophy of Science, 58*(2), 168–184. https://doi.org/10.1086/289610
Nicholson, D. J. (2014). The return of the organism as a fundamental explanatory concept in biology. *Philosophy Compass, 9*(5), 347–359. https://doi.org/10.1111/phc3.12128

Okasha, S. (2011). Biological ontology and hierarchical organization: A defence of rank freedom. In B. Calcott & K. Sterelny (Eds.), *The major transitions in evolution revisited* (pp. 53–64). MIT Press.

Pradeu, T. (2010). What is an organism? An immunological answer. *History and Philosophy of the Life Sciences, 32*(2–3), 247–267.

Pradeu, T. (2016). Organisms or biological individuals? Combining physiological and evolutionary individuality. *Biology and Philosophy, 31*(6), 797–817. https://doi.org/10.1007/s10539-016-9551-1

Psillos, S. (2004). A glimpse of the secret connexion: Harmonizing mechanisms with counterfactuals. *Perspectives on Science, 12*(3), 288–319. https://doi.org/10.1162/1063614042795426

Queller, D. C., & Strassmann, J. E. (2009). Beyond society: The evolution of organismality. *Philosophical Transactions of the Royal Society of London. Series B, Biological Sciences, 364*(1533), 3143–3155. https://doi.org/10.1098/rstb.2009.0095

Rainey, P. B., & Kerr, B. (2010). Cheats as first propagules: A new hypothesis for the evolution of individuality during the transition from single cells to multicellularity. *BioEssays: News and Reviews in Molecular, Cellular and Developmental Biology, 32*(10), 872–880. https://doi.org/10.1002/bies.201000039

Ruiz-Mirazo, K., Etxeberria, A., Moreno, A., & Ibáñez, J. (2000). Organisms and their place in biology. *Theory in Biosciences, 119*(3), 209–233. https://doi.org/10.1078/1431-7613-00018

Ruiz-Mirazo, K., Peretó, J., & Moreno, A. (2004). A universal definition of life: Autonomy and open-ended evolution. *Origins of Life and Evolution of the Biosphere: The Journal of the International Society for the Study of the Origin of Life, 34*(3), 323–346. https://doi.org/10.1023/b:orig.0000016440.53346.dc

Salmon, W. C. (1984). *Scientific explanation and the causal structure of the world.* Princeton University Press.

Santelices, B. (1999). How many kinds of individual are there? *Trends in Ecology & Evolution, 14*(4), 152–155. https://doi.org/10.1016/s0169-5347(98)01519-5

Schlosser, G. (1998). Self-re-production and functionality. *Synthese, 116*(3), 303–354. https://doi.org/10.1023/a:1005073307193

Shapiro, J. A. (2011). *Evolution: A view from the 21st century.* FT Press Science.

Smith, M., & Szathmáry, E. (1995). *The major transitions in evolution.* Oxford University Press.

Sober, E. (1991). Organisms, individuals, and units of selection. In A. I. Tauber (Ed.), *Organism and the origins of self. Boston studies in the philosophy of science* (Vol. 129, pp. 275–296). Springer. https://doi.org/10.1007/978-94-011-3406-4_13

Sousa, R. d. (2005). Biological individuality. *Croatian Journal of Philosophy, 5*(2), 195–218.

Spencer, H. (1864). *The principles of biology.* Williams and Norgate.

Theis, K. R., Dheilly, N. M., Klassen, J. L., Brucker, R. M., Baines, J. F., Bosch, T. C. G., Cryan, J. F., et al. (2016). Getting the hologenome concept right: An eco-evolutionary framework for hosts and their microbiomes. *mSystems, 1*(2), e00028-16. https://doi.org/10.1128/mSystems.00028-16

Wilson, R. A. (2004). *Genes and the agents of life: The individual in the fragile sciences biology.* Cambridge University Press.

Wimsatt, W. C. (1986). Forms of aggregativity. In A. Donagan, A. N. Perovich, & M. V. Wedin (Eds.), *Human nature and natural knowledge: Essays presented to Marjorie Grene on the occasion of her seventy-fifth birthday* (pp. 259–291). Springer. https://doi.org/10.1007/978-94-009-5349-9_14

Wimsatt, W. C. (1997). Aggregativity: Reductive heuristics for finding emergence. *Philosophy of Science, 64*(4), 372–384. https://doi.org/10.1086/392615

Wolvekamp, H. P. (1966). The concept of the organism as an integrated whole. *Dialectica, 20*(2), 196–214. https://doi.org/10.1111/j.1746-8361.1966.tb00516.x

Woodward, J. F. (2001). Law and explanation in biology: Invariance is the kind of stability that matters. *Philosophy of Science, 68*(1), 1–20. https://doi.org/10.1086/392863

Woodward, J. F. (2002). What is a mechanism? A counterfactual account. *Philosophy of Science, 69*(S3), 366–377. https://doi.org/10.1086/341859

Woodward, J. F. (2003). *Making things happen: A theory of causal explanation.* Oxford University Press.

Wright, L. (1973). Functions. *Philosophical Review, 82*(2), 139–168. https://doi.org/10.2307/2183766

Zilber-Rosenberg, I., & Rosenberg, E. (2008). Role of microorganisms in the evolution of animals and plants: The hologenome theory of evolution. *FEMS Microbiology Reviews, 32*(5), 723–735. https://doi.org/10.1111/j.1574-6976.2008.00123.x

Zylstra, U. (1992). Living things as hierarchically organized structures. *Synthese, 91*(1–2), 111–133. https://doi.org/10.1007/bf00484972

Chapter 2
Biological Autonomy and Functional Integration

Abstract This chapter explores the concept of biological autonomy as interpreted through Maturana and Varela's autopoietic framework and the organisational account. It examines the type of functional integration necessary for a biological system to achieve autonomy and teleology. The chapter argues that autonomy requires the mechanistic integration of functions contributing to self-maintenance and agential capacities. This integration demands a specific organisation of the system's components, allowing it to establish its own norms. Consequently, an autonomous system exhibits teleology determined by its internal organisation (internal teleology), rather than by an external agent (external teleology). This distinction between internal and external teleology differentiates organisms from machines. In organisms, functional integration fosters autonomy and internal teleology, whereas in machines, it results in a quasi-autonomous system whose structure and function depend on human social contexts.

Keywords Functional integration · Autopoiesis · Biological autonomy · Organisational framework · Teleology

2.1 Introduction

The previous chapter has demonstrated that functional and mechanistic integration are fundamental conceptual tools for understanding the physiological and evolutionary dimensions of individuals. Organisms are characterised not only as physiological and evolutionary units but also as autonomous systems.

What exactly is autonomy? On the one hand, the term intuitively suggests an independence of biological systems from their surroundings. On the other hand, living beings are not isolated from their environment and constantly interact with it. This dialectic is clearly exemplified by the process of feeding: organisms undertake metabolic activities that are not causally determined by the environment; yet, metabolism necessitates that organisms interact with their environment to obtain nutrients.

G. Militello, *Functional Integration*, History, Philosophy and Theory of the Life Sciences 38, https://doi.org/10.1007/978-3-031-94744-5_2

Philosophers have proposed various, not mutually exclusive, meanings of biological autonomy: the ability of human beings to perform intentional acts (e.g. Collier, 1999; Enç, 2003; Gallagher, 2023; Lumer & Nannini, 2016); the capacity of an organism to self-maintain, act, and transform its environment (e.g. Bickhard, 2000; Moreno & Etxeberria, 2005; Moreno & Mossio, 2015); and the foundation of the goal-directed nature of living beings, also known as teleology (Moreno & Mossio, 2015; Mossio et al., 2009; Mossio & Bich, 2017). All these interpretations of biological autonomy necessitate the formulation of a concept of functional integration. For instance, intentionality can arise if certain areas of the brain are functionally connected and physiologically integrated. Similarly, an organism's self-maintenance and action are properties of the entire system that emerge due to the functional integration among its component parts.

This chapter aims to explore the type of functional integration required for a biological system to be autonomous and teleological. Sections 2.2 and 2.3 examine two key philosophical accounts of biological autonomy: the autopoietic framework and the organisational account, respectively. In both cases, to study the relationship between biological autonomy and functional integration, special attention will be devoted to the mechanisms and biological dimensions whose integration produces a biologically autonomous organisation. Section 2.4 discusses the relationship between autonomy, functional integration, and teleology.

I argue that for a system to be autonomous, it requires mechanistic integration between the functions that contribute to its self-maintenance and agential capacities. This integration necessitates a specific organisation of the system's components, enabling the system to establish its own norms. In doing so, an autonomous system exhibits a teleology determined by its internal organisation (internal teleology), rather than by an external agent (external teleology). This distinction between internal and external teleology is the primary difference between organisms and machines. Both systems require a certain level of functional interdependence. However, in organisms, functional integration leads to autonomy and internal teleology, whereas in machines, it results in a quasi-autonomous system whose structure and function depend on a human social context.

2.2 Living Beings as Autopoietic Systems

The term *autopoiesis* was first introduced in the 1970s by Maturana and Varela (Maturana & Varela, 1973; Varela et al., 1974) to describe a system capable of producing (*poiesis*) its organisation by itself (*auto*). An autopoietic system is defined as "a network of production of components which (i) participate recursively in the same network of productions of components which produced these components, and (ii) realize the network of productions as a unity in the space in which the components exist" (Varela et al., 1974, p. 188). This definition highlights key features of autopoiesis. Firstly, the self-production of the system results from a network of component parts that are interconnected. Secondly, the mutual dependence of these components within the network establishes a cohesive and material unit with clear spatial boundaries.

Maturana and Varela consider the *cell* a paradigmatic example of an autopoietic system because it comprises a network of biochemical reactions, the interactions of which generate and sustain the network itself, rendering the cell a cohesive unit. A cell is autopoietic because its topological and functional organisation is "continuously realized under permanent turnover of matter, regardless of its changes in form and specificity of its constitutive chemical reactions" (Varela et al., 1974, p. 188). This implies that, although the component parts (i.e. cellular molecules) change, the organisation of the whole (i.e. the cell) is maintained over time.

This insight is crucial as it underscores a fundamental aspect of autopoietic systems: the organisation of the whole must remain invariant and stable, even though the component parts can undergo transformation. Accordingly, "if the network of productions of components which define the organisation is disrupted, the unity disintegrates" (Varela et al., 1974, p. 188).

Compared to cells, *human-made machines* are not autopoietic systems. Although they consist of component parts organised to form a unit, "the organisation of a machine does not specify the properties of the components […], it only specifies the relations which these must generate to constitute the machine or system as a unity" (Maturana & Varela, 1980, p. 77). Unlike living beings, the properties of the parts and the function(s) of a machine are not determined by its internal organisation but by an external agent, namely, the engineer who designed it. Furthermore, the use of a machine is defined by the social context in which it operates and "belongs to our description of the machine in a context wider than the machine itself" (Maturana & Varela, 1980, pp. 77–78).

In contrast, *autopoietic*—or "living"—*machines* are homeostatic systems that continuously produce their own organisation and renew their component parts. The design of an autopoietic machine is not the result of an external agent but is instead driven by the internal functional organisation of its component parts. Consequently, only autopoietic machines are:

1. *Autonomous*, because they "subordinate all changes to the maintenance of their own organisation" (Maturana & Varela, 1980, p. 80).
2. *Individual*, as they "maintain an identity which is independent of their interactions with an observer" (Maturana & Varela, 1980, p. 80).
3. *Unities*, since "their operations specify their own boundaries in the process of self-production" (Maturana & Varela, 1980, p. 81).
4. *Homeostatic*, possessing internal mechanisms that compensate for external perturbations, thereby preserving their organisation over time.

Thus, the concept of autopoiesis is considered a "necessary and sufficient [condition] to characterise the organisation of living beings" (Maturana & Varela, 1980, p. 82).

Both human-made and autopoietic machines can be explained in terms of their functions; however, they are organised differently, which explains why some systems are autopoietic while others are not. This raises an important question: what type of integration is necessary for a set of functions to constitute an autopoietic system? Although Maturana and Varela do not explicitly use the term "functional integration", they do provide insights into the different types of causal interdependencies in human-made and autopoietic machines. These observations help to distinguish the distinct forms of functional integration that differentiate these systems.

Human-made machines feature functions that are organised as a straightforward concatenation of processes, enabling them to perform one or more tasks. These functions neither generate the component parts nor shape the architecture of the machine. In contrast, autopoietic machines are characterised by functions that are interdependent and organised according to three key relations: *constitution*, *specification*, and *order*.

Biological functions in autopoietic machines constitute the "topology in which the autopoiesis is realised" (Maturana & Varela, 1980, p. 88). They are specifically organised and ordered to facilitate biological activities. For instance, in a cell—an exemplar of an autopoietic machine—functions are intricately arranged. Macromolecules such as nucleic acids, proteins, carbohydrates, and fatty acids play roles in constitutive relations, such as the production of cell membranes, which delineate the physical boundaries of the cell. These macromolecules also exhibit specification relations that "determine the identity (properties) of the components of the autopoietic organisation" (Maturana & Varela, 1980, p. 91). This is accomplished through nucleic acids (DNA and RNA) and proteins (enzymes) that synthesise all the molecular components of the cell. Additionally, macromolecules govern the temporal order of cellular processes, exemplified by enzymes that regulate the rate of biochemical reactions.

The autopoietic framework provides valuable insights for understanding living beings. Maturana and Varela emphasise the role of organisation and the recursive, self-producing nature of life forms. This perspective aligns with Aristotle's notion of the *formal cause*, which highlights the primacy of an organism's overall architecture—its form and essence—over the individual mechanisms performed by its component parts, akin to the efficient cause. In this view, a living being is autopoietic because its components are arranged in a specific way that allows them to perform mechanisms contributing to the system's self-production.

Furthermore, Maturana and Varela regard the cell as the quintessential example of an autopoietic system. It is seen not only as the fundamental unit of life but also as a model embodying the core characteristics of living systems. By focusing on the cell, they provide a theoretical foundation for understanding more complex forms of life, such as multicellularity, based on the principles demonstrated by cellular organisation.

Nevertheless, Maturana and Varela's framework has notable limitations. While the concept of autopoiesis is closely tied to biological individuality, it does not incorporate the concepts of physiological and evolutionary individuality, as discussed in Sect. 1.4. Their framework lacks an analysis of how physiological dimensions are integrated to form a cohesive biological system capable of evolving as a unit of selection.

In their characterisation of autopoietic systems, Maturana and Varela prioritise self-production over reproduction and evolution. They argue that "reproduction requires the existence of a unity to be reproduced; evolution requires reproduction and the possibility of change" (Maturana & Varela, 1980, p. 96). Consequently, reproduction and evolution are not considered in their definition of living systems. According to their view, "living systems are defined as unities by their autopoiesis" (Maturana & Varela, 1980, p. 96), which implies that self-production takes logical and biological precedence over the processes of reproduction and evolution.

Although Maturana and Varela emphasise the role of mechanisms in understanding both human-made and autopoietic machines, their framework does not address how these mechanisms are integrated to enable the machines to function effectively. For instance, while the biological mechanisms underlying metabolism are crucial for describing autopoietic machines, they argue that "energetic and thermodynamic considerations would necessarily enter into the analysis of how the components are physically constituted; [therefore] these considerations do not enter into the characterization of the autopoietic organization" (Maturana & Varela, 1980, p. 89). This approach suggests a limitation in their framework: by excluding such considerations, they imply that the conceptualisation of autopoietic organisation is independent of the integration of its component mechanisms. In reality, understanding how these mechanisms work together is essential for a comprehensive view of autopoietic systems.

Another problematic aspect of the autopoietic framework is its rejection of biological teleology. The concept of autopoiesis focuses solely on the recursive and self-producing nature of living beings, without incorporating teleological notions, which Maturana and Varela dismiss as relevant only to human actions rather than to biological systems (Maturana & Varela, 1980, p. 85). However, as will be discussed in Sect. 2.4, the organisation of living beings permits the consideration of goal-directed structures, which challenges and relativises Maturana and Varela's dismissal of teleology.

A final critical aspect of the autopoietic framework is its treatment of biological autonomy. Autonomy is portrayed merely as a corollary of autopoiesis, confined to the internal dynamics of self-creation.[1] In this view, autonomy is defined as the system's ability to control the flow of energy and matter to sustain itself. However, the framework does not adequately address the relationship between the system and its environment in defining biological autonomy. This omission has sparked a vigorous debate regarding the nature of biological autonomy, the details of which are explored in the following section.

[1] As Varela pointed out, "the phenomenology of an autopoietic system is the phenomenology of autonomy: all changes of state (internal relations) in the system that take place without disintegration are changes in autopoiesis which perpetuate autopoiesis" (Varela et al., 1974, p. 192).

2.3 The Autonomy of Biological Organisations

Among the philosophical accounts of autonomy,[2] the organisational framework holds a prominent position (Moreno & Mossio, 2015; Mossio et al., 2009; Mossio et al., 2016). This approach conceptualises autonomy based on the specific biological and physico-chemical characteristics—referred to as "organisational features"—of organisms. Although there is some continuity with Maturana and Varela's work, the organisational framework introduces theoretical elements that set it apart from the autopoietic model. These include the role of constraints and their closure in explaining living systems, the distinction between the constitutive and interactive dimensions of autonomy, and the close relationship between biological teleology and autonomy. This section will explore these aspects in detail, focusing on their connection to the concept of functional integration.

2.3.1 Constraints and Closure of Constraints

The concept of *constraint*, originating from classical mechanics, refers to any entity that limits the degrees of freedom of a system without being affected by it (Pattee, 1973). For example, consider the movement of the arm: it is constrained by the shoulder joint, which connects the humerus to the scapula, and the elbow joint, which links the humerus to the radius and ulna of the forearm. These joints not only restrict the arm's range of motion but are also essential for its movement. In this way, constraints, by limiting movement possibilities, paradoxically create a form of freedom (Pattee, 1973, p. 97). In biological systems, constraints are organised *hierarchically*, establishing a hierarchy of control. Higher levels impose specific, dynamic constraints on the movements of lower levels. For instance, the collection of subunits at a higher structural level can also constrain the motion of individual subunits at a lower level (Pattee, 1973, p. 93).

Building on Pattee's framework, Montévil and Mossio (2015) delve into the concept of "constraint" within biological systems, offering a refined definition. They first differentiate between *processes* and *constraints*. Processes encompass the full range of physico-chemical changes within a biological system, involving the alteration, consumption, or constitution of relevant entities. In contrast, constraints refer to entities that influence these physico-chemical processes, regulating their spatial and temporal dynamics without being altered by them. An important characteristic

[2] An important contribution to the conceptualisation of biological autonomy is Bernd Rosslenbroich's "*Biological Autonomy: A New Framework for Understanding Evolution*" (Rosslenbroich, 2014). In this book, Rosslenbroich explores key features of biological autonomy and examines its connections to both microevolution and macroevolution. Unlike some other works that approach biological autonomy from philosophical or organisational perspectives, Rosslenbroich's analysis is grounded in an evolutionary context, offering a fresh viewpoint on how autonomy shapes evolutionary processes.

of constraints is their *context-dependence*; they are tied to specific processes occurring at particular time scales.[3] For instance, an enzyme catalysing a biochemical reaction exemplifies a biological constraint: it affects the rate of a physico-chemical process and alters the spatial configuration of substrates without being impacted by these changes.

Biological systems are open thermodynamic systems that maintain a state far from equilibrium by continuously exchanging energy and matter with their environment. Constraints manage this thermodynamic flow without being influenced by it and without "transmitting energy and/or matter to the process" (Montévil & Mossio, 2015, p. 184). Constraints do not operate in isolation; rather, they interact with one another across various time scales, simultaneously constraining and being constrained. This condition is referred to as the "*closure of constraints*" and describes "an organisation in which each constraint is involved in at least two distinct dependence relationships; in other words, each constraint functions as both a generative and a dependent constraint" (Montévil & Mossio, 2015, p. 186).[4]

Consider enzymes as an example. Enzymes constrain biochemical reactions, yet they are themselves constrained by messenger RNA (mRNA) and ribosomes. Enzymes, being proteins, are synthesised by mRNA and ribosomes. Conversely, mRNA and ribosomes are produced through biochemical reactions catalysed by enzymes, creating a mutual dependence. Thus, enzymes, mRNA, and ribosomes operate within a regime of closure of constraints.

Montévil and Mossio (2015) propose that biological systems exhibit a tendency towards closure, a concept that can be examined by assessing their spatial boundaries. This tendency serves as "a measure of the degree of organisational integration of organisms, as well as an operational tool for delineating their boundaries, even when they are functionally dependent on each other" (Montévil & Mossio, 2015, p. 189). According to the authors, this tendency towards closure also has a

[3] The authors formally define "constraint" as follows: "Given a process A⟶B (A becomes B), C is a constraint on A⟶B, at a specific time scale τ, if and only if the following two conditions are fulfilled: I The situations A⟶B and A_c⟶ B_c (i.e., A⟶B under the influence of C) are not, as far as B is concerned, symmetric at a time scale τ. […]. II A temporal asymmetry is associated with all aspects of $C_{A \longrightarrow B}$ with respect to the process A_c⟶B_c, at time scale τ" (Moreno & Mossio, 2015, p. 182).

[4] A more formal definition of closure is provided by the authors as follows: "A set of constraint *C* realises overall closure if, for each constraint C_i belonging to *C:*

1. C_i depends directly on at least one other constraint belonging to *C* (C_i is dependent).
2. There is at least one other constraint C_j belonging to *C* which depends on C_i (C_i is generative).

A set *C* which realises overall closure also realises strict closure if it meets the following additional condition:

3. *C* cannot be split into two closed sets" (Montévil & Mossio, 2015, p. 186).

normative dimension, as it establishes the criteria for identifying relevant biological entities and their interactions at appropriate time scales.

Montévil and Mossio's concept of the "organisational integration of organisms" closely links the closure of constraints with functional integration. However, these concepts are not synonymous. Closure pertains to the mutual interdependence of entities acting as constraints, whereas functional integration refers to the interdependence of specific functional dimensions within living beings that collectively define them as individuals. While closure of constraints provides the foundation for functional integration, it is not the same as functional integration itself. The closure of constraints underpins functional integration by ensuring that biological capacities and functions depend on the interrelated constraints that regulate the flow of energy and matter within an organism.

In other words, the closure of constraints is concerned with differentiating biological entities from non-biological entities, while functional integration focuses on distinguishing individual biological entities from collective biological organisations. As discussed in Sect. 1.4, the concept of an individual is more specific than that of a biological entity. For instance, viruses, bacterial colonies, and insect colonies are biological entities but do not necessarily qualify as individuals. The distinction between functional integration and closure of constraints is crucial, as functional integration cannot simply be equated with the coordination of biological functions, which alone does not constitute a biological individual. For example, while the heart's beating, blood pumping, and central venous pressure are coordinated functions, these alone do not define a human being as an individual. Instead, a human being is defined as an individual when the cardiovascular system, along with all other organ systems, is functionally integrated to form a cohesive physiological and evolutionary unit.

2.3.2 *Constitutive Autonomy*

Closure of constraints is a crucial component in the theoretical framework of biological autonomy, as it elucidates how a biological system achieves *self-maintenance* and *self-determination*. In the organisational account, these capabilities are described as "constitutive autonomy" because an autonomous entity maintains its structure and function over time (self-maintenance) through its internal regulations rather than external intervention (self-determination). This constitutive dimension is underpinned by the closure of constraints across three fundamental processes: metabolism, gene expression, and biological regulation (see Moreno & Mossio, 2015, chap. 1). These aspects, detailed below, represent "an emergent closure of constraints that [...] grounds functionality" (Moreno & Mossio, 2015, p. 89).

The capacity for self-maintenance and self-determination is deeply intertwined with *thermodynamics*. As previously noted, biological systems are open and dissipative, constantly channelling energy and matter flows to remain far from thermodynamic equilibrium. *Metabolism* encompasses the processes by which an organism converts energy to synthesise molecular components (anabolism) or to degrade these molecules to release energy for work (catabolism) (Moreno & Mossio, 2015; Moreno & Ruiz-Mirazo, 1999). Metabolic reactions involve both endergonic and exergonic transformations: endergonic reactions are non-spontaneous and absorb energy, whereas exergonic reactions are spontaneous and release energy. The production of work relies on the coupling of endergonic and exergonic reactions, while heat is generated as a by-product of exergonic transformations[5] (Kauffman, 1986, 2000).

Metabolic reactions occur at the cellular level and are organised into pathways—sequences of enzyme-catalysed reactions that convert substrates into final products.[6] These pathways rely on two essential types of constraints: enzymes and energy intermediates. Enzymes modulate the rate of metabolic reactions and facilitate the coupling of exergonic and endergonic reactions, ensuring their synchronisation. Energy intermediates, such as NAD^+/NADH and $NADP^+$/NADPH, enable exergonic reactions to drive endergonic ones (Moreno & Ruiz-Mirazo, 1999). Together, enzymes and energy intermediates provide the necessary energy for the system to perform work and carry out all biological functions.

The second dimension of constitutive autonomy is *gene expression*, which refers to the processes by which genetic information encoded in DNA is converted into functional products, such as RNA or proteins. According to Francis Crick's central dogma of molecular biology (Crick, 1958), gene expression involves three main steps: (1) DNA replication by DNA polymerase, (2) transcription of DNA into messenger RNA (mRNA) by RNA polymerase, and (3) translation of mRNA into proteins by tRNA and associated synthetases. DNA and RNA polymerases, as well as tRNA synthetases, are enzymes that constrain the biochemical reactions involved in gene expression, thereby regulating the flow of energy, matter, and information. This process is crucial for the self-maintenance and self-determination of an autonomous living being, as it produces macromolecules—RNAs and proteins—that are essential for fundamental cellular functions, such as metabolism, signalling, and the cell cycle. These macromolecules enable a cell, and by extension, a living organism, to self-produce and self-replace its components, supporting its overall autonomy.

A fundamental aspect of constitutive autonomy is a biological system's ability to compensate for variations, whether these arise internally or externally. In the nineteenth century, Claude Bernard ([1865] 2013) highlighted that living beings maintain an internal milieu that remains stable despite fluctuations. Although living systems constantly exchange energy and matter with their environment, their functional organisation maintains a state of meta-stability through mechanisms that control perturbations. This concept was further developed by Walter Cannon (1929), who introduced the term "homeostasis" to describe a biological system's capacity

[5] Whereas work designates an ordered form of energy, heat is associated with disorder and the degradation of energy transformation (see Moreno & Mossio, 2015, chap. 1).

[6] The idea that metabolism consists of metabolic pathways was also conceptualised by Stuart Kauffman (1986), who coined the expression "autocatalytic set" to designate a series of enzymes that constrain a chemical reaction.

to regulate physiological variables—such as pH, temperature, and the concentrations of sodium, potassium, and calcium—within a narrow range. This regulation ensures that internal conditions remain steady, thus supporting the system's overall stability and autonomy.

The specific way a biological system compensates for variations and maintains homeostasis is through *regulatory processes*. Regulation, the third fundamental dimension of constitutive autonomy, involves second-order constraints that act on the first-order constraints of metabolism and gene expression (Bich et al., 2016; Moreno & Mossio, 2015). Unlike constitutive constraints, which maintain closure under stable conditions, regulatory constraints are crucial for re-establishing closure when it is disrupted. Biological regulation encompasses a range of biochemical mechanisms, such as enzyme phosphorylation and proteins that control gene expression. These regulatory mechanisms modulate metabolic and genetic processes, ensuring the proper functioning of these systems. By doing so, regulation plays a vital role in the self-maintenance and self-determination of a living being (see Chap. 3 for more details).

In light of the above discussion, two distinct senses of "functional integration" emerge from the organisational account of constitutive autonomy.

Firstly, functional integration can be understood as the result of a *closure of constraints* within each of the three dimensions of constitutive autonomy. For instance, metabolism is functionally integrated in that a metabolic network comprises a complex array of enzymes and energy intermediates, all of which are interconnected to regulate biochemical transformations. Similarly, gene expression relies on a network of proteins—including enzymes, transcription factors, and growth factors—that are interdependent in their roles to either activate or inhibit gene expression. Finally, enzyme phosphorylation and gene regulation are intricately integrated to coordinate the control of both metabolism and gene expression. This integration ensures that each dimension of constitutive autonomy supports the overall stability and functionality of the biological system.

A second sense of functional integration pertains to *the interdependence among metabolism, gene expression, and regulation*, which collectively produce the constitutive autonomy of organisms. Metabolism depends on gene expression to synthesise enzymes and other proteins required for biochemical reactions and regulatory processes. Conversely, gene expression relies on metabolic energy to drive the synthesis of RNA and proteins necessary for these processes. Additionally, regulatory mechanisms are fuelled by metabolic energy and involve proteins synthesised through gene expression. The organisational framework emphasises this second sense of functional integration by illustrating how the first-order constraints of metabolism and gene expression interact with the second-order constraints of regulatory processes. However, it does not fully elucidate the mechanisms that facilitate this integration at the most fundamental level of living beings, such as cells. This issue is explored in detail in Chap. 3, which addresses the intricate relationship between constitutive autonomy and functional integration.

2.3.3 Interactive Autonomy

The philosophical literature often uses the term "agency" to denote interactive autonomy, which is the capacity of a system to act (Schlosser, 2019). However, this broad definition complicates the ontological clarity of agency for two main reasons. First, it applies equally to both artificial and biological systems, which can blur the distinctions between them; for example, nanomachines and robots, designed to influence their environment, might be considered agents in a similar manner to organisms. Second, agency manifests differently across various forms of life, from unicellular to multicellular organisms, each exhibiting unique modes of action. This diversity can lead to multiple interpretations and potential ambiguities, as "agency" might be used homonymously with varying significations.

The polysemy of "agency" is illustrated by various contemporary philosophical perspectives. Some authors propose minimal criteria for agency applicable to both simple biological entities and artificial systems. For instance, Dennett (1996) considers self-replicating molecules, such as DNA and RNA, as minimal agents because they represent fundamental forms of life that perform actions. Similarly, other theorists suggest that minimal agency can be identified in simple movements resulting from chemical reactions and self-assembly processes, exemplified by self-propelling oil droplets (Horibe et al., 2011) or nanorobots (Lagzi, 2013). Additionally, Burge (2009, 2010) links primitive agency to the motility of living organisms, involving coordinated behaviours executed by the entire organism.

Other philosophers have associated agency with intentional action, often linking it to mind and consciousness and thus highlighting it as a feature predominantly characteristic of humans. Some theorists argue that agency arises from the causal relationship between mental states—such as beliefs, desires, and intentions—and actions (e.g. Bishop, 1989; Davidson, 1963; Dretske, 1988; Enç, 2003; Goldman, 1970; Mele, 1992). In contrast, alternative views suggest that intentional agency might not require mental representations like desires but could instead be grounded in basic biological drives that lack representational or propositional content (Sterelny, 2001, chap. 11).

In contrast to previous accounts, the organisational approach to agency emphasises its intrinsic connection with autonomy, leading to the term *interactive autonomy*. This framework defines agency through the concept of *minimal agency*, which encompasses a set of biological features shared across all agents, from the simplest unicellular organisms to the most complex multicellular animals (Moreno, 2018; Moreno & Etxeberria, 2005; Moreno & Mossio, 2015).

A minimal agent is characterised by four key features. First, an agent acts upon its environment, exerting causal influence and imposing its own norms on its surroundings.[7] This indicates interactive autonomy, where the agent interacts with the environment rather than being solely acted upon by it. However, this does not imply

[7] The capacity to impose its own norms on the environment is what makes a living being intrinsically teleological as I discuss in more detail in Sect. 2.4.

complete freedom from environmental constraints; rather, the environment shapes and limits the range of possible actions (see Moreno & Mossio, 2015, chap. 4). Second, the constitutive and interactive dimensions of autonomy are mutually dependent and functionally integrated. Interaction with the environment necessitates the ability to manage thermodynamic flows, while metabolic capacities depend on the system's interaction with and modification by its surroundings. Third, agency involves adaptive behaviour, enabling an organism to alter its actions in response to external stimuli. This adaptability requires regulation akin to metabolism and gene expression to adjust according to environmental changes. Lastly, agency is attributed to the entire system rather than to individual components. For instance, while we do not consider an arm alone as an agent, the movement of the arm as part of the whole body contributes to defining the agency of a human being.

Chemotaxis, which refers to the movement of an organism in response to chemical stimuli, is cited by Moreno and Mossio as a prime example of minimal agency.[8] For instance, in the bacterium *Escherichia coli*, chemotaxis involves the rotation of flagella in response to environmental chemicals. This process enables *E. coli* to actively modify its surroundings by interacting with its environment and altering its boundary conditions significantly (Moreno & Mossio, 2015, p. 98). Chemotaxis depends on metabolic processes, as it results from the coordination between membrane receptors and motor mechanisms, which are mediated by cellular metabolic pathways (Moreno & Mossio, 2015, p. 97).

At the same time, chemotaxis supports the self-maintenance of bacteria by facilitating the uptake of sugar, a crucial substrate for their metabolism. Moreover, it exemplifies adaptive agency, as the rotation of the flagella is regulated internally in response to varying sugar concentrations (Barandiaran & Moreno, 2008). Finally, chemotaxis is a characteristic of the entire organism rather than its individual parts, as it involves the integrated functioning of various subsystems, including flagella, metabolism, and regulatory mechanisms.

In chemotaxis, two levels of functional integration underpin this type of agency. First, there is the integration between the mechanisms responsible for sensing nutrients, such as signal transduction mediated by transmembrane receptors, and those controlling flagellum rotation, specifically the Che proteins. Second, chemotaxis is integrated with the broader framework of constitutive autonomy: the rotation of the flagella is regulated through protein phosphorylation and is energetically supported by metabolic processes. This dual level of integration demonstrates how chemotaxis functions both within its immediate biochemical context and in relation to the organism's overall self-maintenance and adaptability.

Beyond minimal agency, more complex forms of interactive autonomy have evolved throughout the history of life. A notable development is the emergence of

[8] It is worth noting that motility is *not* the only form of minimal agency. In the bacterial world, other kinds of minimal agency include gene exchange among bacteria (lateral gene transfer) or the regulation of intercellular signalling based on cell population density (quorum sensing). In the eukaryotic world, fungi interact with their environment through the release of spores, and plants through spores, seeds, and the formation of roots that affect soil morphology.

the nervous system, with primitive versions found in invertebrates such as cephalopods and cnidarians. This advancement introduced two significant innovations. First, there was a dynamical decoupling between agency and metabolism. While the nervous system remains energetically supported by metabolic processes, its biophysical specificity, high connectivity, and electrochemical dynamics afford it a degree of independence from the underlying metabolic organisation (Moreno & Mossio, 2015, p. 175). Second, the nervous system enabled faster and more flexible movements. Neurons facilitate rapid transmission of stimuli across the body, allowing for synchronous movement of multiple motile parts of the organism.

In invertebrates, a single nervous system regulates both physiological functions and motility. In contrast, vertebrates possess two distinct nervous subsystems: the autonomic and somatic nervous systems.[9] The autonomic nervous system manages non-voluntary control of internal organs, while the somatic nervous system oversees voluntary body movements. The differentiation between these systems necessitated advanced coordination facilitated by new brain structures, such as the thalamocortical and limbic systems. These structures not only integrated the autonomic and somatic nervous systems but also enabled the emergence of primitive forms of consciousness, involving feedback loops between memory of values and current perceptual categorisation (Edelman, 1987; Edelman & Tononi, 2000).

The organisation of the nervous systems in invertebrates and vertebrates sheds light on the functional integration required for more complex forms of agency. In invertebrates, where the nervous system is not highly differentiated, motility and sensation are closely intertwined, forming a continuous feedback loop. This integration allows for direct interaction between sensory inputs and motor responses. In contrast, vertebrates have distinct autonomic and somatic nervous systems: the autonomic system regulates non-voluntary functions, while the somatic system controls voluntary movements. This distinction leads to a degree of decoupling between motility and sensation, but they remain functionally integrated through the thalamocortical and limbic systems. The thalamocortical system manages sensorimotor tasks, while the limbic system facilitates communication of emotional and informational flows between the brain and body. Despite the functional decoupling from metabolism, there is an essential integration between the sensorimotor system and metabolic processes. The metabolic system supplies energy to the neuro-muscular system, sustaining its function, while also being influenced by it through the neuroendocrine system and behavioural outputs, such as the exertion of effort until physiological limits are reached (Moreno & Mossio, 2015, p. 174).

The organisational analysis of agency in both minimal forms of life and metazoans highlights two crucial dimensions of functional integration essential for agency: the integration of mechanisms underlying sensorimotor coupling, and the integration of agential capacities with constitutive dimensions such as metabolism and

[9] The differentiation and complexification of the nervous system in vertebrates paved the way for the emergence of primitive forms of consciousness. For more details, see Christensen (2007), Moreno and Etxeberria (2005), Barandiaran and Moreno (2008), Moreno and Mossio (2015, chap. 7).

regulatory processes. While these dimensions elucidate the relationship between constitutive and interactive autonomy, the organisational approach has predominantly focused on case studies involving fully developed organisms, such as bacterial chemotaxis or the nervous systems of metazoans. Chapter 7 will explore how major evolutionary transitions, such as the origin of eukaryotic cells, lead to significant changes in agential capacities and pose philosophical questions, including: What biological transformations enabled the shift from one type of agency to another? What role does motility play in evolutionary transitions? Can motility be preserved while evolving into new functional capacities?

2.4 Autonomy, Teleology, and Functional Integration

The characterisation of living beings as autopoietic (Sect. 2.2) and autonomous (Sect. 2.3) systems highlights a fundamental distinction between natural biological systems and artificial machines: the former can self-determine and transform their environment, while the latter cannot. Constitutive and interactive autonomy arise from biological laws and evolutionary behaviours inherited by organisms. The organisational framework (Moreno & Mossio, 2015; Mossio et al., 2009; Mossio & Bich, 2017) uses the term *norm* to describe the internal laws that underpin an organism's autonomy. This concept of norm implies that living beings pursue biological goals driven by their internal organisation rather than external forces. This raises several important questions: In what sense can an autonomous organisation be considered goal-directed or teleological? Is teleology essential for understanding autonomy? What type of functional integration is necessary for an organism to exhibit teleological behaviour?

The concept of teleology, which refers to explaining phenomena based on their purpose or end goal (*telos*), has been a focal point of philosophical debate concerning biological functions and organisms. Some critics argue that teleology is a human construct, unsuitable for describing the objective reality of living beings, and that it improperly suggests that future effects can shape present traits (Dawkins, 2015; Mayr, 1998). In contrast, the adaptationist programme, a significant approach within evolutionary biology, supports a teleological view by interpreting traits as adaptations shaped by natural selection to enhance evolutionary success. This perspective argues that many biological features can be understood in terms of their functional roles and adaptive advantages (Ayala, 1970; Kitcher, 1993; Millikan, 1989; Neander, 1991). The adaptationist programme can be seen as integrating teleology into evolutionary explanations, where traits are considered to have been selected for their adaptive purposes. However, this view is also subject to critique, as it may overlook non-adaptive explanations such as by-products or developmental constraints. In molecular biology, teleological notions are often reframed in terms of feedback mechanisms and selection pressures, highlighting the complex interplay between adaptation and function (Monod, 1970).

The philosopher Immanuel Kant was one of the first to identify a close connection between teleology, autonomy, and the organisation of organisms. In the *Critique of Judgement* ([1790] 2005), particularly in the First Introduction (§6), Kant presents two ways of describing natural systems: mechanistically and teleologically. Mechanistic descriptions explain natural phenomena by analysing the interactions and functions of their component parts, while teleological descriptions view natural systems as wholes with specific purposes or goals. For instance, the behaviour of plants and animals can be understood either through the mechanisms of their constituent parts or as coherent systems with inherent objectives.

Kant distinguishes between two types of judgements regarding natural systems: determining judgements and reflective judgements. Determining judgements uncover the objective relations and efficient causes within a system. In contrast, reflective judgements, which encompass aesthetic and teleological judgements, reveal a subjective relationship between nature and the human mind.[10] Kant argues that teleological judgements, while they do not reveal objective causal relations, are crucial for understanding the organisation of living systems as purposive wholes. Thus, Kant's framework helps bridge the gap between mechanistic explanations and the teleological perspectives necessary for comprehending the autonomy and complex organisation of biological entities.

According to Kant, a natural purpose is intrinsic to a thing if it is both its own cause and effect. This internal teleology hinges on the concepts of individual and species. Through reproduction, individuals generate offspring that are identical to themselves, thus perpetuating the species. Within a species, an individual is both the effect of its parents and the cause of its progeny, enabling the species to endure despite variations among individual organisms (§64). Additionally, the parts of an individual, though produced by the individual, contribute to its own existence and self-maintenance (§64).

Kant emphasises the recursive nature of organisms, where component parts not only sustain the whole but also perpetuate its existence through reproduction. He describes living beings as organised products of nature in which "everything is the purpose and reciprocally also a means" (§66). This notion underscores that biological organisms are not mere products of chance but embody an internal teleology, where the parts and the whole are interdependent in maintaining and reproducing the organism.

The situation contrasts sharply with artifacts, where the cause and form of their production are not inherent to the artifact itself but rather depend on an external designer. For instance, "in a clock, a cog cannot produce another clock, nor can a clock produce other clocks by using (or organising) other materials" (§65). Kant highlights that in artifacts, neither the parts can reproduce themselves nor can the whole artifact generate identical entities. Artifacts and man-made machines lack the self-producing and self-maintaining qualities of living systems, and thus cannot be

[10] The subjective relation of the human mind with natural objects allows humans to treat natural systems as if they were art objects. Finalistic judgements form the basis for the aesthetic evaluation of natural objects and the contemplation of beauty and the sublime.

considered autopoietic or autonomous. This type of purposiveness is termed external teleology.

Building on Kant's arguments, the organisational approach contends that constitutive and interactive autonomy are intrinsically linked to teleology because "on the one hand, the causal regime of a self-maintaining system provides a naturalised grounding for the teleological dimension. [...] On the other hand, self-maintenance grounds normativity" (Moreno & Mossio, 2015, p. 70). In this view, the existence of a trait within a self-maintaining system can be understood in terms of its current contribution to the system's self-maintenance (Mossio et al., 2009; Mossio & Bich, 2017). For instance, when asked "What is the purpose of the heart?" One might respond, "The goal of the heart is to pump blood, as this activity supports the self-maintenance of the body". Additionally, self-maintenance establishes the norms for an organism's behaviour, since living beings are regulated by physiological parameters (e.g. heart rate, blood pressure, body temperature) that must be maintained within specific ranges to ensure survival. Furthermore, agency, or interactive autonomy, also embodies teleological aspects as actions are directed towards goals that contribute to the system's self-maintenance. Since both constitutive and interactive autonomy rely on the concept of closure of constraints (see Sect. 2.3.1), this concept is considered a foundational element of biological teleology and normativity.

The autopoietic framework, in contrast, argues that teleology should be avoided when characterising biological systems. Maturana and Varela reject the possibility that teleology can be naturalised or possess an objective nature. They assert that teleology is solely a construct of the human mind, having a subjective nature. As they state, "purposes or aims [...] are not features of the organization of any machine (allo- or autopoietic); these notions belong to the domain of our discourse about our actions, that is, they belong to the domain of descriptions" (Maturana & Varela, 1980, p. 85).

These opposing views on biological teleology raise several questions: Can we identify dimensions that need to be integrated to produce a teleological organisation? Can we distinguish internal and external teleology based on different types of functional integration? The autopoietic framework answers these questions negatively, viewing teleology as a useless concept for understanding autopoietic machines. Conversely, the organisational account provides a positive answer: since autonomy grounds normativity, biological teleology relies on integrating functions (or constraints) that ensure (1) the existence of constitutive and interactive autonomy and (2) their functional interdependence. A biological system establishes its own norms and exhibits internal teleology by integrating the functional dimensions underlying constitutive autonomy (e.g. metabolism, gene expression, regulation), agency (e.g. communication, motility, sensorimotor coupling), and the mechanisms that make them interdependent.

When examining machines, can we identify a different type of functional integration that grounds external teleology? Compared to living beings, man-made machines exhibit an important similarity and a key difference. Similar to living beings, the component parts of a machine are assembled to maintain the overall structure in a meta-stable state and constrain an energy input to perform work, realise a systemic function, and contribute to self-maintenance (Militello & Moreno, 2018). However, unlike organisms, their self-maintenance capacity depends on (1) a specific design created by humans and (2) the current use made by humans.

For example, a car's structure and functionality depend on the design and assembly of its component parts. The self-maintenance of a car is ensured as long as a human provides it with fuel and ensures that all the parts (e.g. wheels, engine components) are working correctly and are functionally integrated. Thus, even in machines, functional integration underlies self-maintenance and teleology.

Nevertheless, the main difference from organisms is that the norms of a machine are determined by human design and usage context. Therefore, human intervention has ontological priority over the teleology of machines compared to their internal organisation and functional integration. The autopoietic account asserts that teleology is a subjective way of describing biological phenomena and does not belong to the objective nature of autopoietic systems. Conversely, the organisational framework provides a naturalistic and objective basis for biological teleology. In this context, both internal and external teleology require a functionally integrated organisation, underscoring the close link between functional integration, teleology, and autonomy.

2.5 Concluding Remarks

This chapter has examined the concept of autonomy in biological systems and its relationship with functional integration. Autonomy is typically interpreted as the capacity of living beings to self-maintain and be (relatively) independent from their environment. I have argued that these capacities require the functional integration of specific biological functions.

Maturana and Varela's concept of an autopoietic system emphasises self-production as a fundamental aspect of understanding the autonomy of living beings. Autopoiesis relies on the intimate connection between the parts and the whole: the component parts of an organism engage in mechanisms (e.g. metabolism, gene expression) that produce the component parts and allow the overall system to self-maintain as a whole. Simultaneously, the architecture of the system enables the component parts to contribute to autopoiesis. This capacity for self-creation and self-production distinguishes living beings from human-made machines.

Although Maturana and Varela do not explicitly use the term "functional integration" or emphasise the link between autopoiesis and functional integration, they highlight that biological functions are concatenated in a way that generates relations of constitution, order, and specification. These causal interdependencies explain how an autopoietic system self-produces, reproduces, and evolves.

Compared to the autopoietic account, the organisational framework emphasises not only the role of self-maintenance but also agency in explaining autonomy: through its actions, an organism imposes its own norms on its environment. I have argued that functional integration is crucial to autonomy in three main senses:

1. *Closure of Constraints*: This is the most basic sense in which biological functions are causally interdependent, ensuring the overall system's autonomy. However, closure of constraints is a general term and does not specify which biological dimensions need to be integrated to generate autonomy.
2. *Constitutive and Interactive Dimensions*: Both dimensions of autonomy result from the integration of specific biological capacities. For example, metabolic pathways, gene expression, and regulatory processes are integrated to ensure the system's constitutive autonomy. Similarly, the integration of communication, motility, and sensory capacities underlies agency.
3. *Mutual Integration of Constitutive and Interactive Autonomy*: Constitutive and interactive autonomy are mutually integrated. Metabolism provides agential capacities with the necessary energy, while agential capacities allow the organism to access the nutrients and resources necessary for metabolism.

The concept of autonomy is closely connected with teleology because the biological processes of organisms are aimed at their self-maintenance and self-determination. While the autopoietic framework dismisses biological teleology as a subjective concept, the organisational account argues that it is possible to provide a naturalised and objective foundation for biological teleology. According to this view, a system is teleological if it establishes its own norms, and these norms contribute to its autonomy. Thus, the functional integration required for biological autonomy also provides an ontological basis for the teleology of living beings.

Equating biological teleology with self-determined norms is epistemically fruitful because it helps distinguish physiological from pathophysiological states. The difference between these states lies in their respective contributions to the norms (i.e. range of physiological values) of the organisms. Consequently, addressing the integration of functions required for physiological unity involves considering the range of biological variables that make the overall system physiologically viable.

The relationship between autonomy, teleology, and functional integration provides new insights into biological functions, mechanisms, and individuality (see Chap. 1). Regarding biological functions, their nature is not only shaped by evolutionary history and their current role in the organism's physiology (Mossio et al., 2009), but also by how they are integrated to ensure the autonomy and norms of the living being.

In terms of mechanistic integration, this chapter suggests that biological mechanisms are organised to ensure both constitutive and interactive autonomy, as well as their mutual interdependence. This integration is crucial for understanding how biological systems maintain their internal coherence while interacting with their environment.

Regarding individuality, the integration of constitutive and interactive autonomy offers a theoretical framework for understanding physiological individuality. A biological system is considered a physiological unit if it can self-maintain and interact effectively with its surroundings. Furthermore, if this system possesses reproductive capacities and the ability to transmit variations to its offspring, it qualifies as an evolutionary individuality.

The notion that individuality relies on a functionally integrated autonomous organisation provides a foundational framework for the subsequent chapters. The organisational perspective on autonomy clarifies the dimensions that need to be integrated to achieve autonomy. However, the authors of this perspective do not offer a precise definition of functional integration, nor do they specify which biological dimensions must be integrated to define a physiological and evolutionary individual.[11]

This book aims to complement the organisational framework by exploring the specific functional dimensions necessary for a biological system to qualify as both a physiological and evolutionary individual. As will be demonstrated in the following chapters, the significant changes occurring during the transition from prokaryotes to eukaryotes can be better understood through the lens of the close relationship between (constitutive and interactive) autonomy and functional integration. Therefore, elucidating the connection between functional integration and autonomy is crucial for understanding how these concepts relate to individuality.

References

Ayala, F. J. (1970). Teleological explanations in evolutionary biology. *Philosophy of Science, 37*(1), 1–15. https://doi.org/10.1086/288276

Barandiaran, X., & Moreno, A. (2008). Adaptivity: From metabolism to behavior. *Adaptive Behavior, 16*(5), 325–344. https://doi.org/10.1177/1059712308093868

Bechtel, W., & Bich, L. (2024). Organisms need mechanisms; mechanisms need organisms. In J. Cordovil, G. Santos, & D. Vecchi (Eds.), *New mechanism. Explanation, emergence and reduction* (History, philosophy, and theory of the life sciences) (Vol. 35, pp. 85–108). Springer.

Bernard, C. ([1865] 2013). Introduction à l'étude de la médecine expérimentale. Flammarion.

Bich, L., & Bechtel, W. (2021). Mechanism, autonomy and biological explanation. *Biology & Philosophy, 36*(6), 53. https://doi.org/10.1007/s10539-021-09829-8

Bich, L., & Bechtel, W. (2022a). Control mechanisms: Explaining the integration and versatility of biological organisms. *Adaptive Behavior, 30*(5), 389–407. https://doi.org/10.1177/10597123221074429

Bich, L., & Bechtel, W. (2022b). Organization needs organization: Understanding integrated control in living organisms. *Studies in History and Philosophy of Science, 93*(June), 96–106. https://doi.org/10.1016/j.shpsa.2022.03.005

Bich, L., Mossio, M., Ruiz-Mirazo, K., & Moreno, A. (2016). Biological regulation: Controlling the system from within. *Biology & Philosophy, 31*(2), 237–265. https://doi.org/10.1007/s10539-015-9497-8

Bickhard, M. H. (2000). Autonomy, function, and representation. *Communication and Cognition-Artificial Intelligence, 17*(3–4), 111–131.

[11] It is worth noting that the recent works of Bich and Bechtel represent a pioneering effort to characterise functional integration from both organisational and mechanistic perspectives (Bechtel & Bich, 2024; Bich & Bechtel, 2021, 2022a, 2022b). They emphasise the concept of "control mechanism", arguing that "one way to conceptualize integration and coordination of control is in terms of a pyramid in which, at each level multiple control mechanisms are subsumed under one, with a single control mechanism at the top" (Bich & Bechtel, 2022b, p. 102).

Bishop, J. (1989). *Natural agency: An essay on the causal theory of action*. Cambridge University Press.
Burge, T. (2009). Primitive agency and natural norms. *Philosophy and Phenomenological Research, 79*(2), 251–278. https://doi.org/10.1111/j.1933-1592.2009.00278.x
Burge, T. (2010). *Origins of objectivity*. Oxford University Press.
Cannon, W. B. (1929). Organization for physiological homeostasis. *Physiological Reviews, 9*(3), 399–431. https://doi.org/10.1152/physrev.1929.9.3.399
Christensen, W. (2007). The evolutionary origins of volition. In D. Ross, D. Spurrett, H. Kincaid, & G. Lynn Stephens (Eds.), *Distributed cognition and the will: Individual volition and social context* (pp. 255–287). MIT Press.
Collier, J. D. (1999). Autonomy in anticipatory systems: Significance for functionality, intentionality and meaning. *AIP Conference Proceedings, 465*(1), 75–82. https://doi.org/10.1063/1.58240
Crick, F. H. (1958). On protein synthesis. *Symposia of the Society for Experimental Biology, 12*, 138–163.
Davidson, D. (1963). Actions, reasons, and causes. *Journal of Philosophy, 60*(23), 685–700. https://doi.org/10.2307/2023177
Dawkins, R. (2015). *The blind watchmaker: Why the evidence of evolution reveals a universe without design*. Norton & Company.
Dennett, D. C. (1996). *Kinds of minds: Toward an understanding of consciousness*. Basic Books.
Dretske, F. (1988). *Explaining behavior: Reasons in a world of causes*. MIT Press.
Edelman, G. M. (1987). *Neural Darwinism: The theory of neuronal group selection*. Basic Books.
Edelman, G. M., & Tononi, G. (2000). *Consciousness: How matter becomes imagination*. Allen Lane.
Enç, B. (2003). Autonomy, the will, and freedom. In B. Enç (Ed.), *How we act: Causes, reasons, and intentions* (pp. 219–234). Oxford University Press. https://doi.org/10.1093/0199256020.003.0008
Gallagher, S. (2023). Introduction to the special issue on relational autonomy and collective intentionality. *Southern Journal of Philosophy, 61*(S1), 1–7. https://doi.org/10.1111/sjp.12535
Goldman, A. I. (1970). *A theory of human action*. Princeton University Press.
Horibe, N., Hanczyc, M. M., & Ikegami, T. (2011). Mode switching and collective behavior in chemical oil droplets. *Entropy, 13*(3), 709–719. https://doi.org/10.3390/e13030709
Kant, I. ([1790] 2005). In N. Walker (Ed.), *Critique of judgement*. Oxford University Press.
Kauffman, S. A. (1986). Autocatalytic sets of proteins. *Journal of Theoretical Biology, 119*(1), 1–24. https://doi.org/10.1016/s0022-5193(86)80047-9
Kauffman, S. A. (2000). *Investigations*. Oxford University Press.
Kitcher, P. (1993). Function and design. *Midwest Studies in Philosophy, 18*(1), 379–397. https://doi.org/10.1111/j.1475-4975.1993.tb00274.x
Lagzi, I. (2013). Chemical robotics — Chemotactic drug carriers. *Central European Journal of Medicine, 8*(4), 377–382. https://doi.org/10.2478/s11536-012-0130-9
Lumer, C., & Nannini, S. (2016). *Intentionality, deliberation and autonomy: The action-theoretic basis of practical philosophy*. Routledge.
Maturana, H. R., & Varela, F. J. (1973). *De máquinas y seres vivos: una teoría sobre la organización biológia*. Editorial Universitaria.
Maturana, H. R., & Varela, F. J. (1980). *Autopoiesis and cognition: The realization of the living* (Boston studies in the philosophy and history of science) (Vol. 42). Springer. https://doi.org/10.1007/978-94-009-8947-4
Mayr, E. (1998). The multiple meanings of "teleological". *History and Philosophy of the Life Sciences, 20*(1), 35–40.
Mele, A. R. (1992). *Springs of action: Understanding intentional behavior*. Oxford University Press.
Militello, G., & Moreno, Á. (2018). Structural and organisational conditions for being a machine. *Biology & Philosophy, 33*(5), 35. https://doi.org/10.1007/s10539-018-9645-z
Millikan, R. G. (1989). In defense of proper functions. *Philosophy of Science, 56*(2), 288–302. https://doi.org/10.1086/289488

Monod, J. (1970). *Le Hasard et La Nécessité*. Editions du Seuil.
Montévil, M., & Mossio, M. (2015). Biological organisation as closure of constraints. *Journal of Theoretical Biology, 372*, 179–191. https://doi.org/10.1016/j.jtbi.2015.02.029
Moreno, A. (2018). On minimal autonomous agency: Natural and artificial. *Complex Systems, 27*(3), 289–313. https://doi.org/10.25088/ComplexSystems.27.3.289
Moreno, A., & Etxeberria, A. (2005). Agency in natural and artificial systems. *Artificial Life, 11*(1–2), 161–175. https://doi.org/10.1162/1064546053278919
Moreno, A., & Mossio, M. (2015). *Biological autonomy: A philosophical and theoretical enquiry*. Springer.
Moreno, A., & Ruiz-Mirazo, K. (1999). Metabolism and the problem of its universalization. *Biosystems, 49*(1), 45–61. https://doi.org/10.1016/S0303-2647(98)00034-3
Mossio, M., & Bich, L. (2017). What makes biological organisation teleological? *Synthese, 194*(4), 1089–1114. https://doi.org/10.1007/s11229-014-0594-z
Mossio, M., Montévil, M., & Longo, G. (2016). Theoretical principles for biology: Organization. *Progress in Biophysics and Molecular Biology, 122*(1), 24–35. https://doi.org/10.1016/j.pbiomolbio.2016.07.005
Mossio, M., Saborido, C., & Moreno, A. (2009). An organizational account of biological functions. *British Journal for the Philosophy of Science, 60*(4), 813–841. https://doi.org/10.1093/bjps/axp036
Neander, K. (1991). Functions as selected effects: The conceptual analyst's defense. *Philosophy of Science, 58*(2), 168–184. https://doi.org/10.1086/289610
Pattee, H. H. (1973). *Hierarchy theory: The challenge of complex systems* (1st ed.). G. Braziller.
Rosslenbroich, B. (2014). *On the origin of autonomy: A new look at the major transitions in evolution*. Springer.
Schlosser, M. (2019). Agency. In E. N. Zalta (Ed.), *Stanford encyclopedia of philosophy*. https://plato.stanford.edu/archives/win2019/entries/agency/
Sterelny, K. (2001). *The evolution of agency and other essays*. Cambridge University Press.
Varela, F. G., Maturana, H. R., & Uribe, R. (1974). Autopoiesis: The organization of living systems, its characterization and a model. *Biosystems, 5*(4), 187–196. https://doi.org/10.1016/0303-2647(74)90031-8

Chapter 3
Biological Regulation and Functional Integration

Abstract Biological regulation is a fundamental characteristic of living organisms, enabling them to satisfy internal physiological needs and adapt to environmental changes. This chapter explores the relationship between biological regulation and functional integration. I argue that there is an organisational closure between the regulatory subsystem (R), the processes responsible for self-maintenance (C), and the signalling subsystem (S). Clarifying the functional integration between R, C, and S has two important explanatory outcomes: firstly, it highlights a fundamental organisational dimension of functional integration shared by both prokaryotes and eukaryotes; secondly, it provides a theoretical basis for understanding the evolution of complexity and the emergence of new forms of functional integration during eukaryogenesis.

Keywords Functional integration · Biological regulation · Constitutive processes · Signalling subsystem · Organisational closure

3.1 Introduction

Philosophy of biology has sparked an extensive debate on the nature of biological functions (see Sect. 1.3) and mechanisms (see Sect. 1.4). It has shown that functions and mechanisms occur in organisms, which are autonomous entities (see Chap. 2). Functions, mechanisms, and autonomy are regulated to satisfy the internal physiological needs of organisms and to respond to environmental variations. The nature of biological regulation is inherently problematic. On the one hand, biology often treats regulation as a clear concept, generally synonymous with *control mechanism* and *feedback mechanism*, which enable a system to maintain stable internal physico-chemical conditions (*homeostasis*). On the other hand, "there is no agreement on what regulation actually is. It remains vague, as if it was a concept without precise meaning and with an ambiguous relationship to other concepts akin to it (such as control, homeostasis, feedback, or adaptation)" (Bich et al., 2016, p. 238).

G. Militello, *Functional Integration*, History, Philosophy and Theory of the Life Sciences 38, https://doi.org/10.1007/978-3-031-94744-5_3

Functional integration requires biological regulation, as "all current living systems employ forms of hierarchical control to modulate the relations between their constitutive subsystems in such a way that they are capable [of] coordinating their basic functions and achieving integration" (Bich, 2018, p. 138). This means that the different functional dimensions of an organism (e.g. metabolism, development, reproduction, sensorimotor processes) control and regulate one another, fostering (hierarchical) dependency relationships among them.

In this chapter, I will explore the concept of biological regulation and its relationship with functional integration. I will examine how the constitutive regime—comprising metabolism, gene expression, and development—interconnects with signalling processes and regulatory processes in unicellular organisms such as bacteria. I will analyse two paradigmatic cases: the control of metabolism and development through the regulation of gene expression, and the control of motility (chemotaxis) through enzyme phosphorylation. The rationale behind selecting this case study is straightforward: bacteria exhibit a minimal—but sufficiently complex—functional integration between the constitutive regime (C), signalling (S), and regulatory (R) processes. The interdependence between these three functional subsystems represents a fundamental form of functional integration in both prokaryotic and eukaryotic cells.

I argue that the functional integration between the regulatory subsystem and the constitutive regime consists of a regime of organisational closure in which R directly constrains (i.e. modulates) C, and C can constrain R through the synthesis of R's components and the activity of S. Indeed, R is endogenously synthesised and regenerated by C and is triggered by S, which is generated by C. Clarifying the functional integration between R, C, and S has two significant explanatory consequences: firstly, it reveals a fundamental organisational dimension of functional integration common to prokaryotes and eukaryotes; secondly, it provides a theoretical basis for understanding the evolution of complexity and the emergence of new forms of functional integration in eukaryogenesis.

This chapter is organised as follows. In Sect. 3.2, I critically review some accounts that have explored the relationship between regulatory subsystems and the constitutive regime. Then, in Sects. 3.3 and 3.4, I analyse how regulatory subsystems and the constitutive regime are functionally integrated by studying the regulation of transcription and chemotaxis in bacteria. In Sect. 3.5, I discuss the organisational role of the signalling system in achieving functional integration between R and C. Finally, Sect. 3.6 offers some concluding remarks.

3.2 Regulation in Biological Organisations

Over the past few decades, the most influential and widespread way to define biological regulation has originated from the cybernetic tradition in the life sciences (Wiener, 1948; Ashby, 1956). Within this framework, biological regulation is often associated with *feedback loops* (see Wolkenhauer & Mesarovic, 2005; El-Samad,

2021; Kiesewetter & Schmiemann, 2022), which are circuits where an output is fed back into its input, either opposing it (negative feedback) or enhancing it (positive feedback). This perspective remains influential today across various biological domains, ranging from molecular biology to physiology.

A major criticism of this account has been put forward by Bich et al. (2016, 2020). They argue that feedback mechanisms not only control certain processes but are also directly controlled by them. For example, a thermostat regulates the system's temperature while simultaneously being influenced by it. In contrast, some regulatory mechanisms control processes without being directly influenced by them. For instance, the attachment of a phosphate group to an enzyme (enzyme phosphorylation) regulates the rate of a metabolic reaction without being directly controlled by the reaction's biochemical products.

This perspective invites a broader understanding of biological regulation as encompassing both feedback and non-feedback mechanisms, recognising that complex networks of interactions often govern biological systems.

Bich et al. (2016) argue that regulatory processes act on the constitutive regime, which refers to the set of biological processes—such as metabolism, gene expression, and development—that enable a biological system to maintain itself and exhibit constitutive autonomy (see Sect. 2.3.2). More precisely, regulatory mechanisms operate on the constraints[1] of the constitutive regime, modulating the conditions under which these processes occur to ensure the system's stability and adaptability. The constitutive regime relies on three types of constraints: *kinetic* constraints, which include enzymes that modulate the rate of anabolic and catabolic reactions; *spatial* constraints, which involve structures such as membranes and intermembrane proteins that selectively control the flow of nutrients and waste products through the cellular boundary; and *template* constraints, which involve DNA and mRNA that control the transformation of amino acid sequences into proteins (Bich et al., 2016; Bich, 2018).

Bich et al. (2016) identify five criteria to define biological regulation and distinguish it from the constitutive regime. First, the entities performing regulatory mechanisms (R) must be *endogenously produced* through gene expression by the constitutive regime (C). Indeed, the proteins performing regulatory functions are intracellularly synthesised through genetic transcription and translation, controlled by template constraints. Secondly, regulatory mechanisms must be *decoupled* from the constitutive regime, meaning that "one or more variables in the regulatory subsystem are not directly dependent on the constitutive regime"[2] (Bich et al., 2016, p. 255). Consequently, the regulatory subsystems exhibit a high degree of freedom

[1] Constraints are material structures that exert a causal influence on a physico-chemical process at a specific time scale without being affected by the process itself (see Sect. 2.3.1).

[2] More specifically, this means that the activities of the regulatory subsystems do not stoichiometrically depend on the production of the regulatory subsystems made by the constitutive regime, although the latter is "responsible for the presence and amount of R [the regulatory subsystems] in the system" (Bich et al., 2016, p. 254).

relative to the constitutive regime, allowing them to operate at different rates and remain "not directly dependent on each other" (Bich et al., 2016, p. 254).

Thirdly, regulatory mechanisms are *activated* by specific signals or perturbations in internal or external conditions "rather than by a change in the concentration of the components in R" (Bich et al., 2016, p. 256). Fourthly, regulatory subsystems must play a *functional role*, aiming "to shift (either reversibly or irreversibly) between distinct constitutive/metabolic regimes C, C', C"... available to the system" (Bich et al., 2016, p. 256), thereby contributing to the maintenance of the system. Finally, the influence of R on C enables C to respond to a range of perturbations and *adapt* to new biological conditions.

A key aspect of biological regulation is its dependence on extracellular and intracellular signals that trigger a response in the entities performing regulatory mechanisms.[3] A signalling process consists of a chemical or physical stimulus (first messenger) that is recognised by a protein (the receptor). The receptor transduces the first messenger into a series of second messengers (a signalling cascade) that trigger or inhibit biological processes. Signals can be amplified when a single signalling molecule elicits a response involving a vast number of molecules. Furthermore, they are specific, as a single receptor cannot respond to contradictory signals by simultaneously up-regulating and down-regulating a process (Karp et al., 2020). Hereinafter, the term *signalling system* will refer to the subsystem consisting of the signal transduction machinery and the extracellular/intracellular signals it produces.

In light of the above, it is apparent that R, C, and S somehow work in an integrated way to allow the cell to self-maintain and respond to environmental perturbations. Nonetheless, the features of this functional integration have not yet been thoroughly investigated. Accordingly, in the following section, I will analyse how signals, regulatory subsystems, and the constitutive regime are functionally connected in bacterial cells.

3.3 Regulatory Mechanisms in Bacteria

Bacteria employ two important regulatory strategies: regulating protein synthesis through gene expression (including transcriptional and translational regulation) and modulating enzyme activity through phosphorylation. Both processes enable bacteria to control metabolism, growth, and motility, allowing them to adapt to environmental changes.

[3] In unicellular organisms (i.e. bacteria, archaea, and protists), signals can be both by-products (i.e. metabolites) produced by intracellular or extracellular metabolic processes and chemical elements (e.g. oxygen, minerals, etc.) that interact with a receptor to transduce the signal, thus triggering a regulatory response. In multicellular eukaryotes, extracellular signals mostly consist of hormones, neurotransmitters, and cytokines produced in specific cells and released into the extracellular space.

The regulation of gene expression and enzymatic activity is initiated by both intracellular and extracellular signals. Intracellular signals include metabolites[4] and certain proteins[5] that are synthesised endogenously by cytosolic enzymes, reflecting the internal state of the cell. These endogenous signals typically regulate metabolic pathways. They can also be induced in response to external stimuli and may bind directly to regulatory proteins, such as at the allosteric sites of transcription factors, to exert their effects.

Exogenous signals provide the cell with information about environmental conditions, such as the presence or absence of nutrients. These signals include metabolites and by-products produced by other cells.[6] Exogenous signals are initially detected by regulatory proteins through signal transduction mechanisms. A common type of bacterial signal transduction machinery is the two-component system,[7] where one protein, a sensor kinase,[8] phosphorylates a second protein called a response regulator, which then triggers a regulatory response. Extracellular signals are also transported into the bacterial cytoplasm via various outer membrane transporters (Martínez-Antonio & Collado-Vides, 2003).

3.3.1 The Regulation of Gene Expression

Bacteria possess the ability to regulate the transcription and translation of proteins to meet their physiological requirements. The initiation of transcription is the most critical and controlled step in bacterial gene expression and is mediated by two primary types of proteins: *sigma factors* and *transcription factors*.[9]

Sigma factors are proteins that facilitate the binding of RNA polymerase (RNAP) to gene promoters,[10] thereby initiating transcription.[11] They are instrumental in orchestrating global changes in gene expression in response to environmental

---

[4] The term "metabolite" refers to a wide range of small molecules that are the intermediate end products of metabolism.

[5] For example, the proteins FNR, SoxR, and OxyR sense changes in the redox state of bacteria (Martínez-Antonio et al., 2006).

[6] A cell can receive information about the environment through environmental cues, which are not produced by other cells but by the environment itself. Examples of environmental cues are chemical elements, changes in temperature, pH, and osmotic pressure.

[7] Bacteria can also exhibit one-component systems, which consist of proteins containing input and output domains but lacking histidine kinase and receiver domains (Ulrich et al., 2005).

[8] Sensor kinases are usually integral membrane proteins that can either be a permanent part of a cell membrane (transmembrane integral proteins) or a protein associated with one side of the membrane (monotopic integral proteins).

[9] Several other proteins, such as small ligands and mRNA molecules, are involved in the regulation of bacterial gene expression.

[10] Gene promoters are DNA sequences where gene transcription begins.

[11] Another class of proteins, anti-sigma factors, binds to sigma factors and inhibits transcriptional activity.

stresses and in coordinating gene expression temporally and spatially (Bervoets & Charlier, 2019, p. 309).

Transcription factors (TFs) typically comprise two domains: a DNA-binding domain and an allosteric site. The DNA-binding domain interacts with DNA promoters,[12] acting either as activators or repressors of transcription initiation. Activators facilitate transcription by engaging with the transcriptional machinery, including RNAP, DNA, and general transcription factors. In contrast, repressors inhibit transcription initiation by binding to a DNA sequence known as the operator, thereby obstructing RNAP binding to the promoter (Collado-Vides et al., 1991).

The allosteric site of transcription factors is the binding site for signalling molecules, such as metabolites, which are transmitted via signal transduction pathways. These interactions induce conformational changes in the transcription factors, thereby modulating their capacity to bind DNA and regulate transcription.

There is a close relationship between transcriptional regulation and the transduction machinery in bacteria. A pertinent example is the phosphotransferase system (PTS), specifically the Enzyme I (EI)—Enzyme II (EII) system in *Escherichia coli* (Fig. 3.1). When glucose is present in the periplasmic space, the EI-EII system facilitates its transport and phosphorylation. The unphosphorylated domain of EII interacts with and inactivates Mlc, a transcriptional repressor of the operator region of the *pts* genes.[13] Consequently, the transcription of *pts* genes, which encode proteins responsible for glucose transport, is initiated. Additionally, when glucose is available, the unphosphorylated EIIA domain can bind to and inactivate LacY, the lactose permease, thereby inhibiting lactose transport (Winkler & Hastings Wilson, 1967).

Conversely, in the absence of glucose, the phosphorylated EII domain activates adenylate cyclase, leading to the production of cyclic adenosine monophosphate (cAMP).[14] This molecule binds to the cAMP receptor protein (Crp, also known as Cap), which then binds to DNA to activate the transcription of the lac operon.[15] Moreover, the phosphorylated EII domain does not bind to Mlc, allowing Mlc to repress the transcription of *pts* genes (Plumbridge, 1998).

Transcriptional regulation controls protein synthesis and is therefore involved in modulating a fundamental aspect of the constitutive regime: *metabolism* (i.e. the increase and decrease of enzyme production). The genetic control of metabolism is well exemplified by carbohydrate metabolism regulation in *E. coli*. Glycolysis and gluconeogenesis are controlled by the combined action of the cAMP-Crp complex

---

[12] The DNA promoter is located near to or in an overlapping position with the binding site for RNAP.

[13] The *pts* is a group of genes involved in the synthesis of the phosphoenolpyruvate-dependent sugar phosphotransferase system (sugar PTS) that catalyses the phosphorylation of sugar substrates concomitantly with their translocation across the cell membrane.

[14] cAMP is a second messenger involved in different biological processes in both bacteria and eukaryotes.

[15] The lac operon is a cluster of genes involved in the transport and metabolism of lactose in bacteria.

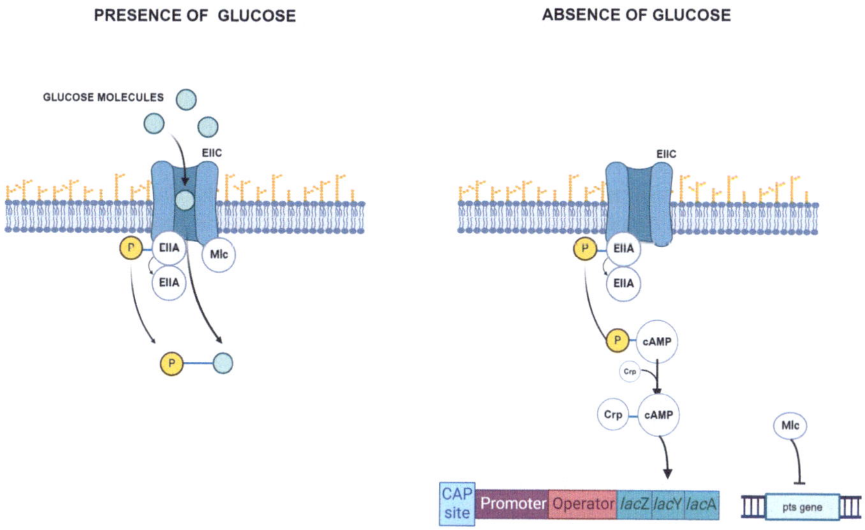

Fig. 3.1 Control of transcriptional regulation in *Escherichia coli* in the presence or absence of glucose. This figure demonstrates how transcriptional regulation in *E. coli* is affected by the availability of glucose. It compares the regulatory mechanisms operational when glucose is present versus when it is absent

and the transcription factor Cra,[16] which are triggered by extracellular glucose signals. When the extracellular concentration of glucose is high, the transcription of genes that synthesise enzymes for carbon uptake and glycolysis (e.g. *ptsHI*, *pfkA*, etc.) is activated, whereas the transcription of genes producing enzymes involved in gluconeogenesis (e.g. *pps* and *pck*), the tricarboxylic acid cycle, the two glyoxylate shunt enzymes, and some electron transport carriers is repressed (Saier & Ramseier, 1996; Shimizu, 2016). The opposite process occurs when there is a low concentration of glucose in the extracellular environment.

Bacterial *growth*, another fundamental dimension of the constitutive regime, is also controlled by transcriptional regulation. Depending on the type and amount of nutrients available, the growth rate of bacteria can vary considerably, being modulated by the regulation of gene expression. For example, when the concentration of amino acids in the environment is low, the alarmone guanosine pentaphosphate (ppgpp) directly binds to the RNA polymerase, repressing the transcription of genes involved in amino acid synthesis and inhibiting the autocatalytic activity of ribosomes, which is essential for bacterial growth (Potrykus et al., 2011; Klumpp & Hwa, 2014).

[16] Cra stands for catabolite repressor/activator protein.

3.3.2 Regulation by Means of Enzyme Phosphorylation

Bacteria can regulate the constitutive regime and interactive capacities not only through the regulation of gene expression but also via post-translational modifications, such as enzyme phosphorylation. This second regulatory strategy is faster than the first because it involves a temporary activation or inactivation of the activity of the constitutive components rather than the activation or inhibition of their production. In this subsection, I examine a classic example of regulation through enzyme phosphorylation: the control of bacterial *chemotaxis*, a type of movement performed by some types of bacteria (Fig. 3.2). Chemotaxis involves modulating the rotatory movement of the bacterial flagellum, which can rotate either clockwise (resulting in tumbling) or counter-clockwise (resulting in running), depending on the concentrations of metabolites in the environment. The regulation of chemotaxis plays a pivotal role in controlling bacterial motility, thereby creating favourable conditions for bacterial metabolism.

Like gene expression regulation, post-translational regulation is triggered by environmental and intracellular signals. In the case of chemotaxis, enzyme phosphorylation is triggered by an increase in repellent concentrations or a decrease in attractant concentrations in the extracellular environment. The transduction of extracellular signals is performed by a two-component system consisting of a sensor kinase (a histidine protein kinase) and a response regulator. The sensor kinase catalyses the transfer of phosphoryl groups from ATP to one of its own histidine residues; the response regulator transfers phosphoryl groups from the kinase phosphohistidine to one of its own aspartic acid residues (Webre et al., 2003). Since most histidine protein kinases are transmembrane proteins, they bind to extracellular signals, leading to the phosphorylation of the response regulator, which then diffuses through the cytoplasm and binds to the motor proteins that regulate flagellar movement.

The histidine protein kinase involved in chemotaxis is CheA,[17] which phosphorylates CheY, the response regulator. Once phosphorylated, CheY binds to a flagellar protein called FliM, inducing a conformational change that modifies the direction of flagellar rotation from counterclockwise to clockwise. This switch in rotation causes the bacterium to tumble. When CheY is dephosphorylated by CheZ, FliM reverts to its original conformation, resulting in the flagellar rotation changing from clockwise to counterclockwise, allowing the bacterium to run.

The regulatory action of the CheA-CheY two-component system is modulated by a set of five receptor transmembrane proteins,[18] which can be methylated or demethylated. Methylation of these receptors inhibits CheA, while demethylation activates CheA, thus fine-tuning the chemotactic response (Webre et al., 2003; Bich & Moreno, 2016).

---

[17] Since CheA is not an integral transmembrane protein, it receives signals and is also regulated by different transmembrane chemotaxis receptors.

[18] The five proteins are: Tar, Tsr, Trg, Tap, and Aer.

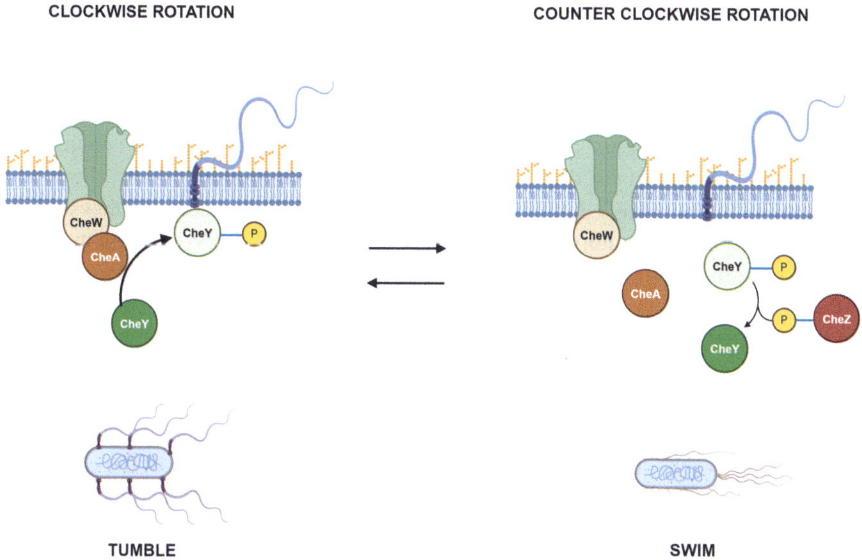

Fig. 3.2 Regulation of chemotaxis through enzyme phosphorylation. This figure illustrates the mechanisms of enzyme phosphorylation involved in chemotaxis. On the left side, the processes associated with clockwise rotation are depicted, while the mechanisms involved in counterclockwise rotation are shown on the right

By adjusting the direction and duration of runs and tumbles, bacteria can navigate toward favourable conditions, such as higher concentrations of nutrients, or away from harmful substances. This regulation through phosphorylation allows bacteria to quickly respond to environmental changes, complementing slower genetic regulatory mechanisms.

3.4 Functional Integration between Constitutive Constraints and the Regulatory Subsystem

The previous section demonstrated that prokaryotes regulate their metabolism, growth, and motility by controlling gene expression and enzymatic activity. The case studies also showed that regulatory mechanisms are triggered by signals, such as the metabolites produced by a bacterium, and by environmental cues, like the glucose present in the surroundings. As pointed out by Bich et al. (2016), constitutive processes are *directly* controlled by regulatory mechanisms and *indirectly* by signals acting on those mechanisms. Thus, there appears to be a direct causal relationship from signals to regulatory and constitutive processes (S → R → C). However, do constitutive processes also play a causal role in influencing S and R? In this section, I will discuss how the constraints of the constitutive regime (i.e.

template, kinetic, and spatial constraints) affect the behaviour of regulatory mechanisms and signals.

There is a reciprocal causal loop between *template constraints* (i.e. DNA and mRNA) and regulatory proteins: regulatory proteins modulate gene expression, while template constraints synthesise and regenerate the second-order constraints. Three important aspects of this causal loop are worth highlighting. First, the structure (and its replacement), but not the functionality, of the regulatory subsystem is controlled by template constraints. For example, the regulatory mechanisms performed by transcription factors do not depend on the genes that synthesise them. Secondly, most transcriptional regulatory proteins regulate genes that are not involved in their own production, so there is no feedback loop between a transcriptional regulatory protein and the gene that synthesises it. For instance, consider the relationship between the global regulator Cra and the genes it regulates. The Cra protein in *E. coli* is a global transcription factor synthesised by the *cra* gene; however, the Cra protein does not constrain the transcription mechanism of the *cra* gene. Instead, it regulates the transcription of genes encoding biosynthetic and oxidative enzymes and sugar catabolism, such as *ppsA*, *fbp*, and *fruB*. Thirdly, transcriptional regulatory proteins are not regulated by first-order constraints since they are capable of self-regulation and regulation among themselves, resulting in a hierarchical network in which we can distinguish between global[19] and local[20] regulators.

Kinetic constraints, such as enzymes, modulate the activity of regulatory subsystems by acting as coactivators or corepressors. For instance, histone acetyltransferases function as coactivators, while histone deacetylases act as corepressors. These enzymes assist regulatory proteins involved in transcriptional regulation. Enzymes like kinases are directly involved in signal transduction systems, which are an essential component of regulatory phenomena. Moreover, enzymes catalyse metabolic processes that produce metabolites, which in turn activate regulatory subsystems[21] as intracellular or extracellular signals.

Spatial constraints, such as porins, membrane receptors, and signal transduction systems, affect regulatory mechanisms by selectively controlling the extracellular signals passing through the outer and inner bacterial membranes. Furthermore, signal transduction systems, such as two-component systems and phosphotransferase systems, initiate kinase cascades that lead to the phosphorylation of response regulator proteins. These proteins play a fundamental role in regulating both gene expression and enzymatic activity.

---

[19] Global regulators not only regulate themselves but also other TFs through feedback loops and complex relationships. Global regulators can work with other global and local regulators to co-regulate the same promoters (Martínez-Antonio & Collado-Vides, 2003).

[20] Local regulators are regulated by global regulators and regulate a single or very few operons (Martínez-Antonio & Collado-Vides, 2003, p. 484).

[21] For example, the enzymes involved in glycolysis (e.g. hexokinase, phosphofructokinase) transform glucose into pyruvate, producing a number of metabolites (e.g. fructose-1,6-bisphosphate and pyruvate) that affect the conformation and regulatory activity of TFs in *E. coli* (Martínez-Antonio & Collado-Vides, 2003).

In summary, the regulation of bacterial transcription and the control of bacterial chemotaxis demonstrate a *bidirectional* relationship between the constraints of constitutive processes (first-order constraints) and regulatory mechanisms (second-order constraints). On the one hand, regulatory mechanisms control the first-order constraints without being reciprocally controlled by them. On the other hand, template, kinetic, and spatial constraints modulate the functionality of regulatory proteins in three key ways. Template constraints are involved in the synthesis and turnover of regulatory proteins through protein synthesis. Spatial constraints, such as transmembrane proteins, can act as regulatory mechanisms by controlling the flow of molecules and signals passing through them. Kinetic constraints contribute to the production of metabolites that act as signals, which, in turn, trigger or inhibit regulatory proteins.

Signals serve as a crucial connecting point between the functions performed by the constitutive constraints and those carried out by the regulatory subsystems. Although signals are produced by metabolic processes, they do not directly constrain these processes. Instead, they constrain regulatory proteins, creating a functional integration between signalling, regulatory, and constitutive processes. The organisational features and systemic physiological consequences of this type of functional integration are discussed in the next section.

3.5 The Organisational Closure between the Constitutive Regime, Regulatory Subsystem, and Signalling Subsystem

The regulation of gene expression and enzyme phosphorylation demonstrates that the constitutive regime, regulatory subsystems, and signalling subsystems are functionally interdependent. This raises the question of whether their underlying constraints give rise to *organisational closure* and what this implies for the overall bacterial organisation.

As pointed out by Montevil and Mossio (Montévil & Mossio, 2015, p. 186), a system is subject to closure if the following conditions are satisfied:

(i) Each constraint C_i of the system *depends directly* on at least another constraint belonging to the system (C_i is dependent).
(ii) Each constraint C_i *generates* at least another constraint C_j belonging to the system (C_i is generative).
(iii) The system cannot be *split* into two closed sets.

Let us examine whether the constraints of the three subsystems—(R), (C), and (S)—satisfy these three requirements of organisational closure.

(i) *Dependency of Constraints*: Regulatory proteins directly influence the behaviour of spatial, template, and kinetic constraints. Regulatory constraints act on template constraints, which in turn modulate enzyme synthesis. Kinetic constraints govern metabolic processes and the production of metabolites, which act as signals that trigger, and therefore constrain, regulatory proteins. Spatial constraints (e.g. transmembrane proteins) regulate the flow of extracellular metabolites between the exterior and interior environments. Thus, the constraints of each subsystem directly depend on other constraints within the system.

(ii) *Generative Nature of Constraints*: The constraints of each subsystem are generative in the sense that they enable "behaviours and outcomes that would otherwise be impossible" (Montévil & Mossio, 2015, p. 183). The regulatory subsystem allows the constitutive regime to adapt to environmental conditions and maintain a stable internal cellular environment. The constitutive regime generates the proteins that make up the regulatory subsystem and some molecules of the signalling subsystem. Finally, the signalling subsystem enables the regulatory subsystem to be activated or inhibited in response to external and internal stimuli.

(iii) *System Unity*: The entire system cannot be divided into two (or more) closed sets, as its proper functioning relies on the mutual interdependence of the three subsystems.

The constraints of the regulatory, constitutive, and signalling subsystems satisfy Montévil and Mossio's requirements and are subject to *organisational closure* (Fig. 3.3). This does not disprove the thesis that regulatory constraints must be not directly dependent (i.e. dynamically decoupled) on the constitutive constraints (Bich & Moreno, 2016; Bich et al., 2016; Bich, 2018). Although regulatory proteins are synthesised by template constraints, their functionality does not hinge on these template constraints but rather on signals. Likewise, the mechanisms performed by regulatory proteins do not depend on the concentrations of metabolic substrates and products (Bich et al., 2016); rather, they depend on how specific metabolites bind to receptors (i.e. allosteric modification), thereby triggering a regulatory response.

Signals and signal transduction machineries (i.e. the *signalling subsystem*) play a fundamental organisational role in the functional integration between the regulatory subsystems and the constitutive regime. In particular, four properties of the signalling subsystem allow it to act globally as an interface between the constitutive regime and the regulatory subsystem. First, the signal transduction machinery can recognise a wide set of molecules (notably metabolites) as signals, *amplifying* them and generating an output that activates a regulatory response. Second, the signal transduction machinery is highly *specific* in recognising signals, ensuring that not all metabolic products activate a regulatory response. Third, the signal transduction machinery *integrates* different signal transduction pathways and regulatory responses. Finally, the signal transduction machinery is a fundamental source of *temporal coordination* in the regulatory response. Let us address each of these four features.

The bacterial signal transduction machinery, such as the two-component system and the phosphotransferase system, typically consists of a transmembrane receptor protein that binds to a ligand to modify a response regulator and produce an output.

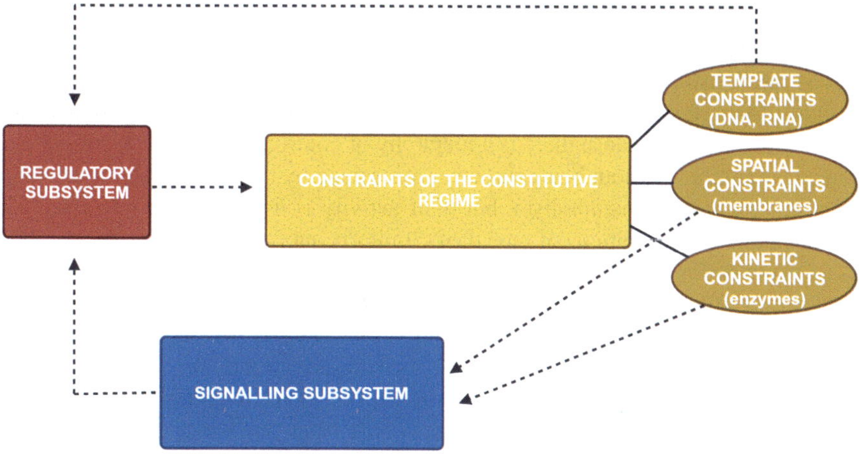

Fig. 3.3 Organisational closure among constitutive regime (C), regulatory subsystem (R), and signalling subsystem (S). This figure illustrates the organisational closure between C, R, and S. It depicts how these subsystems are interrelated and function together within the system

In this basic mechanism, the ligand (e.g. a metabolite) acts as a signal by binding only to specific classes of receptors, inducing a conformational change in these receptors. This change usually leads to the phosphorylation of the response regulator. When a receptor interprets a metabolite as a signal, the signal transduction machinery transforms a direct product of metabolism into a message that can activate a regulatory response. To prevent the distortion of information, especially over long distances, a signal needs to be amplified through signalling cascades.

The faithful transmission of information requires specificity in the interaction between histidine kinases and the response regulators in the two-component system, avoiding the overlapping of signalling pathways (cross-talk). This specificity is achieved through bifunctionality, where histidine kinases either phosphorylate their response regulators or remove a phosphoryl group from the same response regulators in the absence of a stimulus (Salazar & Laub, 2015). By balancing the phosphorylation level of response regulators, histidine kinases prevent cross-talk between different signalling pathways (Groban et al., 2009; Salazar & Laub, 2015). Additionally, the accurate transmission of information is achieved through positive and negative feedback mechanisms that modulate the activity of histidine kinases, controlling the expression levels of their target genes over time.

The activity of signal transduction machineries is modulated by a group of proteins known as "two-component system connectors" (or simply "connectors"), which play an essential role in the coordination and fine-tuning of cellular processes (Mitrophanov & Groisman, 2008). Connectors phosphorylate or dephosphorylate response regulators, thereby modulating the activity of sensor kinases. They facilitate the temporal coordination between signal transduction pathways and regulatory responses. For example, certain connectors, such as RapA and RapE, prevent the simultaneous transcription of two alternative sets of genes (late competence genes and sporulation genes), thus ensuring that distinct regulatory responses are temporally coordinated (Mitrophanov & Groisman, 2008; Gao & Stock, 2017).

The relationship between the regulatory, constitutive, and signalling subsystems reveals a biological architecture in which control is both *hierarchical* and *heterarchical*. This occurs because there is a hierarchy of control in which each subsystem influences another without being directly constrained by it. For example, regulatory subsystems constrain metabolism, but their activity is not directly constrained by the metabolic concentrations of enzymes. Similarly, enzymes influence the synthesis of many endogenous signals (such as metabolites), but their catalytic activity is not directly affected by ligand-receptor interactions. Signals modulate regulatory gene expression and enzyme phosphorylation, but kinase cascades are not directly constrained by them.

It is important to emphasise that within the cell, there is no privileged level of control establishing rigid top-down mechanisms from one subsystem to another. This type of biological organisation has been well characterised by Winning and Bechtel (2018), who argued that subsystems of biological organisations often exhibit local hierarchies where, instead of a strict hierarchy, each subsystem can exert control "over mechanisms at the top of the local hierarchy" (Winning & Bechtel, 2018, p. 299), thus realising a heterarchical control model (Pattee, 1991). Therefore, although regulatory subsystems control metabolic processes, they are, in turn, controlled by the signalling system. As such, it is not possible to consider the regulatory subsystem as the top-level of a threefold control hierarchy, which controls without being controlled.

The closure between the regulatory subsystem, constitutive regime, and signalling subsystems has significant systemic consequences for the bacterial cell. It allows the cell to sense changes in the concentrations of metabolites and other molecules in the internal and external environment, thereby triggering gene regulation or enzyme phosphorylation. This, in turn, controls the protein activity involved in metabolic and developmental processes. The signalling system initiates a regulatory response that switches between different regimes of metabolic and growth processes, regulating them according to physiological requirements, thus substantially contributing to the self-maintenance of the overall cell. Furthermore, the functionality of the signal transduction machinery clearly demonstrates how the regulation of the intracellular environment depends on information about the extracellular environment. The flow of information from the exterior to the interior of the cell is enabled by the signal transduction machinery.

Since signal transduction machineries allow for temporal coordination between variations in metabolite concentrations and regulatory responses, they help cells to synchronise metabolic needs with the activation of regulatory subsystems. This leads to a system capable of responding to internal and external perturbations and changing its mode of behaviour flexibly (Barkai & Leibler, 1997; Kitano, 2004; Klosik et al., 2017).

3.6 Concluding Remarks

In this chapter, I have explored whether and how regulatory subsystems and the constitutive regime are functionally integrated by examining two paradigmatic cases of biological regulation in bacteria: the regulation of gene expression and regulation through enzyme phosphorylation. These two case studies demonstrate that regulatory subsystems directly modulate the action of constraints (i.e. template and enzymatic constraints) by activating or deactivating protein synthesis and by triggering or inhibiting the activity of proteins. In turn, the constraints of the constitutive regime modulate the behaviour of regulatory subsystems in two key ways: firstly, template constraints synthesise the proteins that perform regulatory functions; secondly, they initiate the functionality of regulatory subsystems by producing and controlling the flow of metabolites that act as signals.

I have argued that the signalling system (comprising signals and signal transduction machineries) plays a fundamental organisational role in the functional integration between regulatory subsystems and the constitutive regime. This is because it allows metabolic products to indirectly affect (in the form of signals) the behaviour of regulatory proteins. As a result, the constraints of the regulatory, signalling, and constitutive subsystems are organisationally closed, as they depend on each other.

It has been shown that intracellular signals directly modulate regulatory responses, thereby informing regulatory subsystems about the intracellular metabolic status. In contrast, extracellular signals must be "read" by the signal transduction machinery, which exhibits four important properties: interpretation and amplification of the ligand, specificity of the ligand-receptor binding, integration of signal transduction pathways and regulatory responses, and temporal coordination in the regulatory response. As such, intracellular signals act as an interface between the constitutive regime and the regulatory subsystem, while intercellular signals serve as a connecting point between the extracellular environment and the intracellular milieu.

The closure between signalling, regulatory, and constitutive subsystems has important physiological consequences for the bacterial cell. It enables the regulation of metabolic and growth processes based on information from both the intracellular and extracellular environments. The activities of these three subsystems require temporal coordination, which is facilitated by the signal transduction machinery (Gao & Stock, 2017). Ultimately, the integration of the three subsystems enables the cell to withstand perturbations, thus achieving biological robustness.

Why has this type of functional organisation been evolutionarily successful, and why does it represent the organisational core for achieving a strong form of functional integration in eukaryogenesis? To address this question, we should consider three key aspects: the maintenance of internal dynamical stability of the constitutive processes, the capacity for (co)evolution, and the potential for incorporating new levels of functional complexity.

The integration between the regulatory, constitutive, and signalling subsystems enables the cell to handle internal and external perturbations by providing appropriate gene and enzyme regulation of the constitutive processes (i.e. metabolism and development). As a result, the cell maintains dynamic stability, compensating for the effects of perturbations by adjusting tightly coupled constitutive constraints (Weiss, 1968; Rosen, 1970). Consequently, this kind of biological organisation is more likely to adapt to a wide variety of environmental niches, survive, and differentially reproduce, thus undergoing natural selection.

For the regulatory, constitutive, and signalling subsystems to be evolutionarily successful, they likely co-evolved, such that structural and functional modifications in each subsystem were intimately connected with changes in the other two. For example, the evolution of the signal transduction machinery (notably the two-component system) opened up new cellular functions, including the emergence of new regulatory and metabolic capacities (Perez & Groisman, 2009; Capra & Laub, 2012).

During the transition from prokaryotic to eukaryotic cells, the subsystems must have been capable of incorporating new levels of complexity. The evolution towards more complex and larger signalling systems comes with significant costs: maintaining the accuracy of signals, avoiding unwanted cross-talk, and preserving the fidelity of information flow within the cell becomes more challenging (Laub, 2016). Thus, we can conclude that the evolution from prokaryotic to eukaryotic cells necessitated the development of new forms of signal specificity and recognition.

The closure between the regulatory, constitutive, and signalling subsystems has allowed for increasing complexity in this basic architecture, not only by adding new structures and functions but also by achieving more complex forms of functional integration between the subsystems. As will be explained in the following chapters, the transition from prokaryotic to eukaryotic cells was characterised by the development of new forms of gene regulation (e.g. post-transcriptional, and post-translational modifications), new signal transduction pathways (e.g. the development of complex intracellular signals enabled by the eukaryotic cytoskeleton), and modifications of constitutive processes (e.g. oxidative phosphorylation and the electron transport chain) while maintaining the organisational closure between the regulatory, constitutive, and signalling subsystems present in prokaryotes. This achievement of more complex forms of functional integration between the subsystems enabled the proto-eukaryotic cell to acquire new physiological viability.

References

Ashby, W. R. (1956). *An introduction to cybernetics*. Chapman&Hall.

Barkai, N., & Leibler, S. (1997). Robustness in simple biochemical networks. *Nature, 387*(6636), 913–917. https://doi.org/10.1038/43199

Bervoets, I., & Charlier, D. (2019). Diversity, versatility and complexity of bacterial gene regulation mechanisms: Opportunities and drawbacks for applications in synthetic biology. *FEMS Microbiology Reviews, 43*(3), 304–339. https://doi.org/10.1093/femsre/fuz001

Bich, L. (2018). Robustness and autonomy in biological systems: How regulatory mechanisms enable functional integration, complexity and minimal cognition through the action of second-

order control constraints. In M. Bertolaso, S. Caianiello, & E. Serrelli (Eds.), *Biological robustness. Emerging perspectives from within the life sciences* (pp. 123–147). Springer.
Bich, L., & Moreno, A. (2016). The role of regulation in the origin and synthetic modelling of minimal cognition. *BioSystems, 148*, 12–21. https://doi.org/10.1016/j.biosystems.2015.08.002
Bich, L., Mossio, M., Ruiz-Mirazo, K., & Moreno, A. (2016). Biological regulation: Controlling the system from within. *Biology & Philosophy, 31*(2), 237–265. https://doi.org/10.1007/s10539-015-9497-8
Bich, L., Mossio, M., & Soto, A. M. (2020). Glycemia regulation: From feedback loops to organizational closure. *Frontiers in Physiology, 11*, 69. https://doi.org/10.3389/fphys.2020.00069
Capra, E. J., & Laub, M. T. (2012). Evolution of two-component signal transduction systems. *Annual Review of Microbiology, 66*, 325–347. https://doi.org/10.1146/annurev-micro-092611-150039
Collado-Vides, J., Magasanik, B., & Gralla, J. D. (1991). Control site location and transcriptional regulation in escherichia coli. *Microbiological Reviews, 55*(3), 371–394. https://doi.org/10.1128/mr.55.3.371-394.1991
El-Samad, H. (2021). Biological feedback control-respect the loops. *Cell Systems, 12*(6), 477–487. https://doi.org/10.1016/j.cels.2021.05.004
Gao, R., & Stock, A. M. (2017). Quantitative kinetic analyses of shutting off a two-component system. *mBio, 8*(3), e00412–e00417. https://doi.org/10.1128/mBio.00412-17
Groban, E. S., Clarke, E. J., Salis, H. M., Miller, S. M., & Voigt, C. A. (2009). Kinetic buffering of cross talk between bacterial two-component sensors. *Journal of Molecular Biology, 390*(3), 380–393. https://doi.org/10.1016/j.jmb.2009.05.007
Karp, G., Iwasa, I., & Marshall, W. (2020). *Karp's cell and molecular biology* (9th ed.). Wiley.
Kiesewetter, A., & Schmiemann, P. (2022). Understanding homeostatic regulation: The role of relationships and conditions in feedback loop reasoning. *CBE Life Sciences Education, 21*(3), ar56. https://doi.org/10.1187/cbe.21-04-0092
Kitano, H. (2004). Biological robustness. *Nature Reviews Genetics, 5*(11), 826–837. https://doi.org/10.1038/nrg1471
Klosik, D. F., Grimbs, A., Bornholdt, S., & Hütt, M.-T. (2017). The interdependent network of gene regulation and metabolism is robust where it needs to be. *Nature Communications, 8*(1), 534. https://doi.org/10.1038/s41467-017-00587-4
Klumpp, S., & Hwa, T. (2014). Bacterial growth: Global effects on gene expression, growth feedback and proteome partition. *Current Opinion in Biotechnology, 28*, 96–102. https://doi.org/10.1016/j.copbio.2014.01.001
Laub, M. T. (2016). Keeping signals straight: How cells process information and make decisions. *PLoS Biology, 14*(7), e1002519. https://doi.org/10.1371/journal.pbio.1002519
Martínez-Antonio, A., Janga, S. C., Salgado, H., & Collado-Vides, J. (2006). Internal-sensing machinery directs the activity of the regulatory network in escherichia coli. *Trends in Microbiology, 14*(1), 22–27. https://doi.org/10.1016/j.tim.2005.11.002
Martínez-Antonio, A., & Collado-Vides, J. (2003). Identifying global regulators in transcriptional regulatory networks in bacteria. *Current Opinion in Microbiology, 6*(5), 482–489. https://doi.org/10.1016/j.mib.2003.09.002
Mitrophanov, A. Y., & Groisman, E. A. (2008). Signal integration in bacterial two-component regulatory systems. *Genes & Development, 22*(19), 2601–2611. https://doi.org/10.1101/gad.1700308
Montévil, M., & Mossio, M. (2015). Biological organisation as closure of constraints. *Journal of Theoretical Biology, 372*, 179–191. https://doi.org/10.1016/j.jtbi.2015.02.029
Pattee, H. H. (1991). Measurement-control heterarchical networks in living systems. *International Journal of General Systems, 18*, 213–221.
Perez, J. C., & Groisman, E. A. (2009). Evolution of transcriptional regulatory circuits in bacteria. *Cell, 138*(2), 233–244. https://doi.org/10.1016/j.cell.2009.07.002
Plumbridge, J. (1998). Control of the expression of the manXYZ operon in escherichia coli: Mlc Is a negative regulator of the mannose PTS. *Moleular Microbiology, 27*(2), 369–380. https://doi.org/10.1046/j.1365-2958.1998.00685.x

Potrykus, K., Murphy, H., Philippe, N., & Cashel, M. (2011). ppGpp Is the major source of growth rate control in E. Coli. *Environmental Microbiology, 13*(3), 563–575. https://doi.org/10.1111/j.1462-2920.2010.02357.x

Rosen, R. (1970). *Dynamical system theory in biology: Stability theory and its applications*. Wiley.

Saier, M. H., & Ramseier, T. M. (1996). The catabolite repressor/activator (Cra) protein of enteric bacteria. *Journal of Bacteriology, 178*(12), 3411–3417. https://doi.org/10.1128/jb.178.12.3411-3417.1996

Salazar, M. E., & Laub, M. T. (2015). Temporal and evolutionary dynamics of two-component signaling pathways. *Current Opinion in Microbiology, 24*, 7–14. https://doi.org/10.1016/j.mib.2014.12.003

Shimizu, K. (2016). Metabolic regulation and coordination of the metabolism in bacteria in response to a variety of growth conditions. *Advances in Biochemical Engineering/Biotechnology, 155*, 1–54. https://doi.org/10.1007/10_2015_320

Ulrich, L. E., Koonin, E. V., & Zhulin, I. B. (2005). One-component systems dominate signal transduction in prokaryotes. *Trends in Microbiology, 13*(2), 52–56. https://doi.org/10.1016/j.tim.2004.12.006

Webre, D. J., Wolanin, P. M., & Stock, J. B. (2003). Bacterial chemotaxis. *Current Biology, 13*(2), R47–R49. https://doi.org/10.1016/s0960-9822(02)01424-0

Weiss, P. A. (1968). *Dynamics of development: Experiments and inferences*. Amsterdam Elsevier.

Wiener, N. (1948). *Cybernetics: Or control and communication in the animal and the machine*. Wiley.

Winkler, H. H., & Hastings Wilson, T. (1967). Inhibition of beta-galactoside transport by substrates of the glucose transport system in escherichia coli. *Biochimica Et Biophysica Acta, 135*(5), 1030–1051. https://doi.org/10.1016/0005-2736(67)90073-9

Winning, J., & Bechtel, W. (2018). Rethinking causality in biological and neural mechanisms: Constraints and control. *Minds and Machines, 28*(2). https://doi.org/10.1007/s11023-018-9458-5

Wolkenhauer, O., & Mesarovic, M. (2005). Feedback dynamics and cell function: Why systems biology is called systems biology. *Molecular bioSystems, 1*(1), 14–16. https://doi.org/10.1039/b502088n

Chapter 4
Eukaryogenesis and the Appearance of a New Individuality

Abstract This chapter examines three principal theories of eukaryogenesis and elucidates why the transition from prokaryotes to eukaryotes constitutes a major evolutionary transition, as defined by Maynard-Smith and Szathmáry. It argues that these theories, alongside Maynard-Smith and Szathmáry's framework, offer intriguing insights into this evolutionary process, which remains an enigma. From a philosophical perspective, eukaryogenesis challenges our understanding of how new biological organisations attain a sufficient degree of functional integration to become autonomous, physiological, and evolutionary individuals. Therefore, the concept of functional integration is central to comprehending eukaryogenesis.

Keywords Eukaryogenesis · Major evolutionary transitions · Endosymbiotic models · Autogenous models · Inside-out models · Functional integration

4.1 Introduction

The history of life is characterised by the emergence of new organisms that have evolved from their ancestors through natural selection and adaptation. Biologists Smith and Szathmáry (1995) introduced the term "major evolutionary transitions" to describe critical steps in evolution that led to the appearance of fundamentally new types of organisms. The transition from prokaryotes to eukaryotes is considered one such major transition because it resulted in the emergence of a completely new type of organism with a functionally integrated organisation that enabled the development of autonomous individuals.

This chapter addresses two main questions: How is eukaryogenesis explained in current evolutionary biology, and in what sense is it a major evolutionary transition? How is eukaryogenesis connected to the philosophical issues of biological autonomy and individuality?

To answer these questions, the chapter first examines the theoretical framework of Maynard Smith and Szathmáry, highlighting how eukaryogenesis represents a major transition. It then outlines the key unresolved aspects of this transition and

G. Militello, *Functional Integration*, History, Philosophy and Theory of the Life Sciences 38, https://doi.org/10.1007/978-3-031-94744-5_4

discusses the three leading theories that attempt to explain the shift from prokaryotes to eukaryotes. Finally, it explores how evolutionary theories on eukaryogenesis relate to philosophical discussions on autonomy and individuality.

I argue that it is not currently possible to determine which evolutionary theory best explains eukaryogenesis. Instead, the three theories of eukaryogenesis, along with Maynard Smith and Szathmáry's framework, collectively provide inspiring and intriguing insights into this evolutionary step, which remains a conundrum. Philosophically, I contend that eukaryogenesis, as a major evolutionary transition, raises questions about how a new biological organisation achieved a sufficient degree of functional integration to become autonomous and to function as both a physiological and evolutionary individual. Therefore, the concept of functional integration is central to understanding eukaryogenesis, and this chapter sets the stage for the subsequent chapters of the book.

The chapter is organised as follows: Sect. 4.2 defines the concept of a major evolutionary transition and explains why eukaryogenesis qualifies as such a transition. Section 4.3 addresses the main unresolved issues surrounding eukaryogenesis, highlighting its status as a conundrum. Section 4.4 presents the three leading evolutionary biology theories that attempt to explain eukaryogenesis. Section 4.5 explores the philosophical questions of individuality and autonomy within the context of eukaryogenesis. Section 4.6 provides some concluding remarks.

4.2 Major Evolutionary Transitions and the Place of Eukaryogenesis

The term "major transitions in evolution" was coined in 1995 by biologists John Maynard Smith and Eörs Szathmáry to describe specific periods in the history of life during which biological organisations undergo radical transformations, leading to the emergence of new living entities (Maynard Smith & Szathmáry, 1995). Major transitions in evolution are characterised by an increase in complexity[1] that "depended on a small number of major transitions in the way in which genetic information is transmitted between generations" (Maynard Smith & Szathmáry, 1995, p. 3). Some of these transitions, such as eukaryogenesis and the origin of meiotic sex, occurred uniquely, whereas others, like the origin of multicellularity and animal societies, happened independently multiple times.

[1] The authors acknowledge that defining or measuring complexity is challenging. Therefore, they use "complexity" to refer to "organisational complexity", which encompasses both structural (i.e. morphological) and functional complexity. Maynard-Smith and Szathmáry note that complexity should not be equated with "progress" or "fitness" (see Maynard Smith & Szathmáry, 1995, p. 5). They suggest that genome size and coding DNA might serve as criteria for measuring complexity, but concede that what these data "tell us about structural and functional complexity is very limited" (Maynard Smith & Szathmáry, 1995, p. 6).

Maynard Smith and Szathmáry shift the focus from a molecular and phylogenetic approach to a more holistic and philosophical perspective in evolutionary biology. The central issue is the emergence of new forms of individuality, which entails that "entities that were capable of independent replication before the transition can replicate only as part of a larger whole after it" (Maynard Smith & Szathmáry, 1995, p. 6). This larger entity represents a new kind of individuality and a new type of functionally integrated organisation. The rise of these larger entities is counterbalanced by a selection process acting at the lower level, which tends to disrupt integration at the higher level.[2] In other words, the emergence of new forms of individuality involves resolving the conflict between the new individual and its constituent parts, which were previously independent and autonomous.

There are three main biological processes that favour the maintenance of higher-level entities once they have arisen. The first is the genetic similarity between the units that merge to form the higher organism: new forms of individuality are characterised by components with a high genetic relatedness. For example, multicellular organisms develop from a single fertilised egg, producing genetically identical cells. Secondly, contingent irreversibility means that the transformation of an entity into part of a larger individual is an irreversible process. As Maynard Smith and Szathmáry explain, "if an entity has replicated as part of a larger whole for a long time, it may have lost the capacity for independent replication that it once had" (Maynard Smith & Szathmáry, 1995, p. 9). Thirdly, the maintenance of higher-level entities depends on central control, which entails that the new entity governs its parts so that their behaviours are strictly regulated and constrained by the whole.

The emergence of new, larger entities is characterised by the specialisation of their functional parts. For example, in the transition from prokaryotes to eukaryotes, the division of labour is facilitated by the compartmentalisation of functions into organelles.[3] Furthermore, each evolutionary transition is marked by new methods of transmitting information from one generation to the next. For instance, the origin of template replication of polynucleotide-like molecules has enabled unlimited heredity, while the appearance of multicellularity is characterised by the co-option and expansion of epigenetic mechanisms for processes like cellular differentiation.

Maynard Smith and Szathmáry identified eight major evolutionary transitions in the history of life. The first is the transformation of replicating molecules into populations of molecules within compartments. The second is the emergence of chromosomes from independent replicators. The third is the origin of DNA and the genetic code from RNA. The fourth is the genesis of eukaryotes from prokaryotic cells. The fifth is the development of sexual populations from asexual clones. The sixth is the emergence of eukaryotic multicellular organisms from eukaryotic unicellular organisms (protists). The seventh is the origin of colonies from solitary individuals.

---

[2] For example, the emergence of sexual reproduction is counterbalanced by parthenogenesis, which has a "twofold advantage over a sexual one" (Maynard Smith & Szathmáry, 1995, p. 8).

[3] The authors observe: "in prokaryotes, there is a single cell compartment, whereas in eukaryotes the genetic nucleus and metabolic cytoplasm are separated, and additional organelles evolved, some recruited symbiotically" (Maynard Smith & Szathmáry, 1995, p. 13).

Finally, the eighth transition is the emergence of human societies employing language from societies of primates.

Eukaryogenesis is considered a major evolutionary transition because it involves a dramatic change in the structural and functional organisation of prokaryotic cells, which include bacteria and archaea. The appearance of eukaryotic cells led to an increase in complexity and the emergence of a completely new kind of biological individual (Fig. 4.1). Although the main features of prokaryotes, such as the cell membrane, flagella, ribosomes, cytoplasm, and cytoskeleton, are still present in eukaryotes, they have undergone extensive modifications. For example, the cell membranes of prokaryotes and eukaryotes differ in the enzymes involved in the biosynthesis of phospholipids,[4] resulting in different membrane capacities. Likewise, the components constituting prokaryotic and eukaryotic flagella differ, providing different types of movement.[5] Furthermore, the size[6] and composition[7] of ribosomes, the cell structures responsible for protein synthesis, vary between prokaryotes and eukaryotes. Compared to the prokaryotic cytoplasm, the eukaryotic cytoplasm is compartmentalised into distinct organelles, forming the eukaryotic endomembrane system,[8] which facilitates a variety of new functions not found in

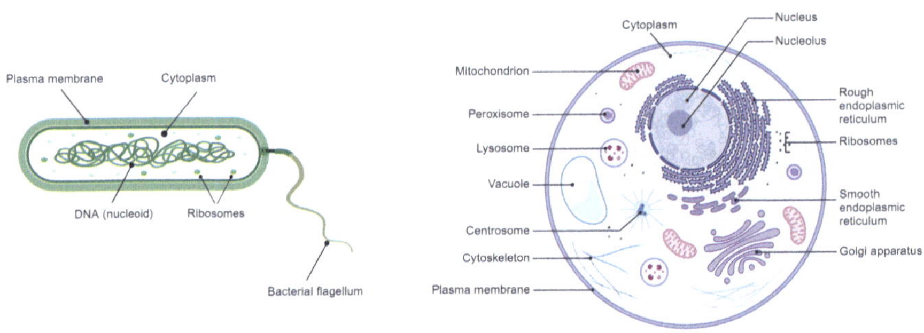

Fig. 4.1 Schematic representation of a prokaryotic cell (on the left) and a eukaryotic cell (on the right). Prokaryotic cells, including bacteria and archaea, are characterised by a simple structure with a plasma membrane that encloses the cell, a flagellum for locomotion, and DNA that is not enclosed within a membrane-bound nucleus. In contrast, eukaryotic cells are more complex, featuring numerous internal membranes, distinct organelles, and a larger overall size. Some unicellular eukaryotes, such as protists, may also possess a flagellum for movement

[4] Specifically, bacteria and eukaryotes utilise glycerol-3-phosphate, while archaea use glycerol-1-phosphate. However, even this distinction has exceptions; for instance, some bacteria have membranes containing mixtures of both types.

[5] The motion in prokaryotes is planar and wave-like, whereas eukaryotes exhibit three-dimensional motion with distinct power and recovery strokes.

[6] Prokaryotes possess 70S ribosomes, each comprising a small (30S) and a large (50S) subunit. In contrast, eukaryotes have 80S ribosomes, consisting of a small (40S) and a large (60S) subunit.

[7] Here, "composition" refers to the number of ribosomal proteins.

[8] The eukaryotic endomembrane system includes the nuclear envelope, endoplasmic reticulum, Golgi apparatus, lysosomes, vacuoles, and vesicles.

prokaryotes.[9] Finally, variations in the composition and organisation of the cytoskeleton[10] give rise to important new eukaryotic features, such as mitosis and meiosis.

Compared to other major evolutionary transitions, eukaryogenesis presents some unique characteristics. The emergence of the eukaryotic cell is closely linked to the engulfment of one prokaryote by another, establishing a long-term symbiotic relationship. This process, known as endosymbiosis, is believed to be the origin of mitochondria and chloroplasts. The emergence of a new form of individuality through symbiosis is a distinguishing feature of eukaryogenesis that is not observed in other evolutionary transitions. Another unique aspect is the challenge of characterising this endosymbiotic relationship: was it a mutualistic relationship, marked by metabolic cooperation between the symbiotic partners, or was it a form of exploitation, in which the host benefited at the expense of the endosymbiont's metabolic capacities? (see Maynard Smith & Szathmáry, 1995, pp. 140–142).

The transition from prokaryotes to eukaryotes can be explained by considering *genetic similarity*, *contingent irreversibility*, *central control*, *specialisation of functional parts*, and new forms of *information transmission*. First, the prokaryotic ancestors of eukaryotic cells were genetically distinct. The endosymbiotic process facilitated a massive transfer of genes from the endosymbiont (the proto-mitochondrion and proto-plastid) to the host, thereby increasing genetic similarity between the two organisms. However, it is important to note that some genetic differences still exist today, as the mitochondrial genome is distinct from the nuclear genome. Regarding contingent irreversibility, the transformation of endosymbionts into mitochondria and plastids was an irreversible process. While there are eukaryotic cells without mitochondria (e.g. the unicellular *Monocercomonoides*), mitochondria have never been observed reverting to the previous stage of endosymbionts.

Eukaryogenesis also involved the establishment of central control, where the host organism developed mechanisms to regulate the behaviour of its endosymbionts and organelles. The specialisation of functional parts is another key feature of eukaryogenesis, as demonstrated by the division of labour among eukaryotic organelles and within the cytoplasm. Finally, the emergence of mitosis introduced new ways of transmitting information from one generation to another. These characteristics will be further examined in the next two sections, which explore the most important hypotheses on the origin of the eukaryotic cell.

---

[9] As discussed in Chap. 6, some bacterial species possess a primitive endomembrane system. However, this is not linked to the emergence of new and more complex biological functions as seen in eukaryotes.

[10] As will be elaborated in Chap. 7, the prokaryotic cytoskeleton mainly consists of a few proteins (FtsZ, MreB, ParM, and crescentin), whereas the eukaryotic cytoskeleton is made up of proteins (actin, myosin, tubulin) arranged into filaments (microfilaments, microtubules, intermediate filaments).

4.3 Eukaryogenesis as a Biological Conundrum

Eukaryogenesis is a significant conundrum, with various scientific hypotheses but no definitive answers. The term "eukaryogenesis" refers to the extended period between the emergence of the first eukaryotic common ancestor (FECA) and the last eukaryotic common ancestor (LECA).[11] The scarcity of fossils from proto-eukaryotic cells and the complete absence of living intermediate forms between prokaryotes and eukaryotes make it extremely challenging to gather conclusive evidence regarding when eukaryotic cells first appeared.

Although still debated, the most likely hypotheses suggest that FECA emerged between 2.7[12] and 1.8[13] billion years ago, while LECA dates back to approximately 1.2[14] billion years ago (Javaux, 2007; Eme et al., 2014).[15] Evolutionary biologists generally agree that eukaryotic cells resulted from the evolutionary transformation of prokaryotes. However, research on eukaryogenesis involves difficult and unresolved questions, which can be summarised as follows: What was the sequence of events leading to the emergence of eukaryotic cells? Which prokaryotic species gave rise to eukaryotic cells? How can we explain the genetic chimerism[16] of eukaryotic cells? I will address these questions in this section.

The first hypothesis regarding the events leading to the formation of eukaryotic cells was proposed by the Russian biologist Konstantin Mereschkowsky in 1910 (Mereschkowski, 1910).[17] Based on his research on lichens, he suggested that the chloroplasts found in algae and plants originated from cyanobacteria that entered another bacterium and were gradually transformed into chloroplasts. He argued that chloroplasts were not newly formed structures, but rather the result of the division of pre-existing chloroplasts. Similarly, he proposed that the nucleus originated through the endosymbiosis between a mycoplasma (the symbiont) and an amoeboid

---

[11] Throughout this chapter and the book, I will use the term "protoeukaryotic cell" to refer to the ancestor of the modern eukaryotic cell, spanning from the period of FECA (First Eukaryotic Common Ancestor) to LECA (Last Eukaryotic Common Ancestor).

[12] The 2.7 billion-year timeframe corresponds to the emergence of the first sterol pathways. Since most prokaryotes do not synthesise sterols, it has been hypothesised that the breakdown products of sterols might serve as a "signature" of early eukaryotic cells (Brocks et al., 1999; Javaux, 2007).

[13] The 1.8 billion-year mark refers to the fossils of *Grypania spiralis*, considered one of the possible earliest ancestors of eukaryotes. Fossils of this organism have been found in iron formations dating from 2.1 to 1.8 billion years ago (Han & Runnegar, 1992).

[14] The 1.2 billion-year mark coincides with the fossils of the red alga *Bangiomorpha pubescens*, which is regarded as the first known sexually reproducing organism with the typical organisation of modern eukaryotic cells (Javaux, 2007).

[15] It is worth noting that the temporal gap between the appearance of the first prokaryotes and FECA is significant, as the earliest prokaryotic fossils date back to 3.45 billion years ago (Zimmer, 2009).

[16] Genetic chimerism refers to the presence of genes in a genome that originate from different species.

[17] An excellent introduction to the history of Mereschkowski's paper can be found in Kowallik and Martin (2021).

plasma (the host).[18] Mereschkowsky suggested that these symbiotic partners were entirely unrelated organisms that arose at different times and under different conditions.

For several decades, Mereschkowsky's endosymbiotic view of eukaryogenesis was dismissed, as the mainstream perspective in evolutionary biology held that eukaryogenesis did not result from endosymbiosis between prokaryotes. Instead, it was believed to be the outcome of small mutations accumulating through natural selection following the divergence of proto-eukaryotes and archaea. In 1967, Lynn Margulis (then known as Lynn Sagan) revisited Mereschkowsky's theory by arguing that eukaryogenesis was the result of a fusion process between different species (Sagan, 1967). This hypothesis was supported by the observation that mitochondria retain their own genes and can replicate, producing new mitochondria. Furthermore, Margulis argued that many mitochondrial genes were of bacterial origin, suggesting a common phylogenetic ancestry between mitochondria and bacteria.

In 1977, Carl Woese and George Fox published pioneering research[19] on the phylogenetic relationships among prokaryotic and eukaryotic cells (Woese & Fox, 1977). They proposed that there were three primitive kingdoms—eubacteria, archaebacteria, and urkaryotes—that independently evolved into the three domains of bacteria, methanogenic bacteria, and eukaryotes, respectively. Regarding the origin of the eukaryotic cell, Woese and Fox suggested that chloroplasts undoubtedly have a "specific eubacterial origin" (Woese & Fox, 1977, p. 5090). However, unlike Margulis, they raised questions about the other organelles and structures, noting that "mitochondria, for example, do not conform well to a 'typically prokaryotic' phenotype" (Woese & Fox, 1977, p. 5090).

The phylogenetic relationship between bacteria, archaea, and eukaryotes is still debated in evolutionary biology and is further complicated by the genetic chimerism of eukaryotic cells. Indeed, eukaryotic nuclear genes have distinct evolutionary origins: some are derived from bacteria, while others come from archaea (see Golding & Gupta, 1995; Rivera & Lake, 2004; Brueckner & Martin, 2020; Nobs et al., 2022). Genes involved in genetic information functions (e.g. replication, transcription, translation, splicing) are derived from archaea, whereas those involved in metabolic functions come from bacteria (Koonin, 2010; Brueckner & Martin, 2020). One hypothesis to explain genetic chimerism is that the ancestor of the eukaryotic cell was an archaeon that established an endosymbiotic relationship with an α-proteobacterium, which later evolved into mitochondria. The massive gene transfer from mitochondria to the proto-nucleus would have resulted in this genetic chimerism (see Lane, 2015, chap. 5).

---

[18] Mycoplasma are now classified as prokaryotes, while amoeboplasma are classified as eukaryotes. The introduction of the terms "prokaryotes" and "eukaryotes" into biological literature dates back to 1925 and became more widespread from the 1960s (Katscher, 2004).

[19] Woese and Fox advanced phylogenetic studies by introducing the use of the Svedberg unit (S) to distinguish and compare ribosomes across species. The 16S Svedberg unit refers to prokaryotic small subunit rRNA, while the 18S unit refers to the equivalent in eukaryotic cytoplasmic ribosomes.

The exact sequence of events leading to the last eukaryotic common ancestor (LECA) remains a topic of debate. Several theoretical models have been proposed, offering a pluralistic and multifaceted view of the main events responsible for eukaryogenesis. We will examine these models in the following section.

4.4 Theoretical Models to Explain Eukaryogenesis

The theoretical models of eukaryogenesis can be divided into three main groups[20] based on the mechanisms involved in the emergence of the eukaryotic nucleus and whether the nucleus appeared before or after the mitochondria.

The first group comprises the *endosymbiotic* models (Lake & Rivera, 1994; Golding & Gupta, 1995; Moreira & Lopez-Garcia, 1998; Margulis et al., 2000; Horiike et al., 2004; Forterre, 2011; López-García & Moreira, 2020), which propose that the nucleus has an endosymbiotic origin. The second category includes the *autogenous* models (Taylor, 1976; Cavalier-Smith, 1988, 2010; Martin & Koonin, 2006), which strongly dispute the endosymbiotic origin of the nucleus, asserting that it was an autogenous process resulting from the invagination of the plasma membrane of the proto-eukaryotic cell. Finally, the third class encompasses the *inside-out* model (Baum & Baum, 2014), according to which the first event in eukaryogenesis was the acquisition of mitochondria. In this model, mitochondria were not endosymbionts, but rather ectosymbionts[21] that were engulfed through protrusions outside the cell wall of the proto-eukaryotic cell. The nucleus is thought to have appeared through an autogenous process following the acquisition of mitochondria.

Interestingly, none of these models offers a definitive account of the temporal sequence of eukaryogenesis: they can consider mitochondria as the initial event, followed by the appearance of the nucleus and all other organelles, or, alternatively, explain the appearance of the nucleus as the earliest event, followed by the development of the endomembranous system and mitochondria/plastids as a later event.

Despite their differences, the endosymbiotic, autogenous, and inside-out models converge on the origin of mitochondria and chloroplasts, as they all acknowledge that these organelles arose through the engulfment of one bacterium by another.[22] Phylogenetic analyses of mitochondria and chloroplasts have revealed that

[20] The theoretical models of eukaryogenesis discussed in this section draw inspiration from Baum and Baum (2014) and Baum (2015), who differentiated between endosymbiotic, autogenous, and inside-out models.

[21] Ectosymbionts (or epibiotic organisms) are those that establish a symbiotic relationship with their host by living on its surface.

[22] The inside-out model posits that mitochondria and chloroplasts originated from ectosymbiosis rather than endosymbiosis, with ectosymbionts being engulfed by the host's external protrusions. Nonetheless, all three models agree on the engulfment of mitochondria and plastids.

mitochondrial genes derive from α-proteobacteria[23] (Ferla et al., 2013), while plastids originate from cyanobacteria (Moore et al., 2019). Recent phylogenetic studies (Gabaldón, 2021) suggest that Asgard archaea may be descendants of the ancestral archaeal host that formed a symbiosis with the ancestral α-proteobacteria, leading to the origin of the mitochondrion.

The reason why proto-eukaryotic cells engulfed α-proteobacteria has been convincingly explained in terms of *metabolic mutualism*: both organisms established a symbiotic relationship that provided a mutual metabolic advantage. In 1967, Margulis (Sagan, 1967) suggested that the symbiotic process was driven by anaerobic bacteria recognising an evolutionary opportunity to thrive in an oxygen-rich environment. Only those capable of adapting to and harnessing oxygen could successfully inhabit this new ecological niche.[24] As such, heterotrophs "were forced to eat organic matter produced by photosynthetic or chemoautotrophic processes" (Sagan, 1967, p. 228). In 1998, two independent and similar metabolic theories for the origin of mitochondria were presented: the *hydrogen hypothesis* and the *syntrophic hypothesis* (Fig. 4.2).

The hydrogen hypothesis was proposed by Martin and Müller (1998),[25] while the syntrophic hypothesis was proposed by Moreira and Lopez-Garcia (1998). Both hypotheses share the idea that eukaryogenesis resulted from an anaerobic metabolic (syntrophic) relationship between an archaeon, which produced methane (CH_4), and a proteobacterium, which produced hydrogen (H_2) and carbon dioxide (CO_2). Initially, the archaeon—the precursor to the eukaryotic cell—began using the H_2 and CO_2 produced by the proteobacterium (the precursor to the mitochondrion) to generate CH_4.[26] This metabolic relationship was the driving force behind the establishment of an endosymbiotic relationship, benefiting the host by allowing it to increase its metabolic rate through a hydrogen source. Subsequently, a dramatic metabolic shift occurred, leading the host to cease methane production in favour of using oxygen.

---

[23] Alphaproteobacteria are gram-negative bacteria lacking peptidoglycan. This group includes several subclasses, one of which is Rickettsia. Recent phylogenetic evidence suggests that the ancestor of mitochondria is a sister group to the entire alpha-proteobacterial cluster (Muñoz-Gómez et al., 2022).

[24] Margulis's claim is implicitly flawed: a strictly anaerobic host and an aerobic bacterium would not naturally meet to form a symbiotic relationship because they inhabit different environments. Physical proximity is essential for symbiosis, and these microorganisms would not typically encounter each other in the same location. Today, it is suggested that an anaerobic organism, such as an archaeon, could combine with a facultative bacterium. This bacterium, capable of both anaerobic metabolism (for the initial encounter) and aerobic metabolism (allowing the consortium to thrive in oxygenated environments), would enable the partnership to establish itself successfully.

[25] The hydrogen hypothesis was deeply influenced by the discovery of hydrogenosomes. These are membrane-enclosed organelles that produce energy (i.e. ATP molecules) in some anaerobic ciliates, flagellates, and fungi. They were discovered by Lindmark and Müller (1973) and their function is the biosynthesis of iron-sulphur clusters.

[26] Martin and Müller argue that this scenario is realistic, because methanogenic archaea rely on water and carbon dioxide for producing energy and biomass.

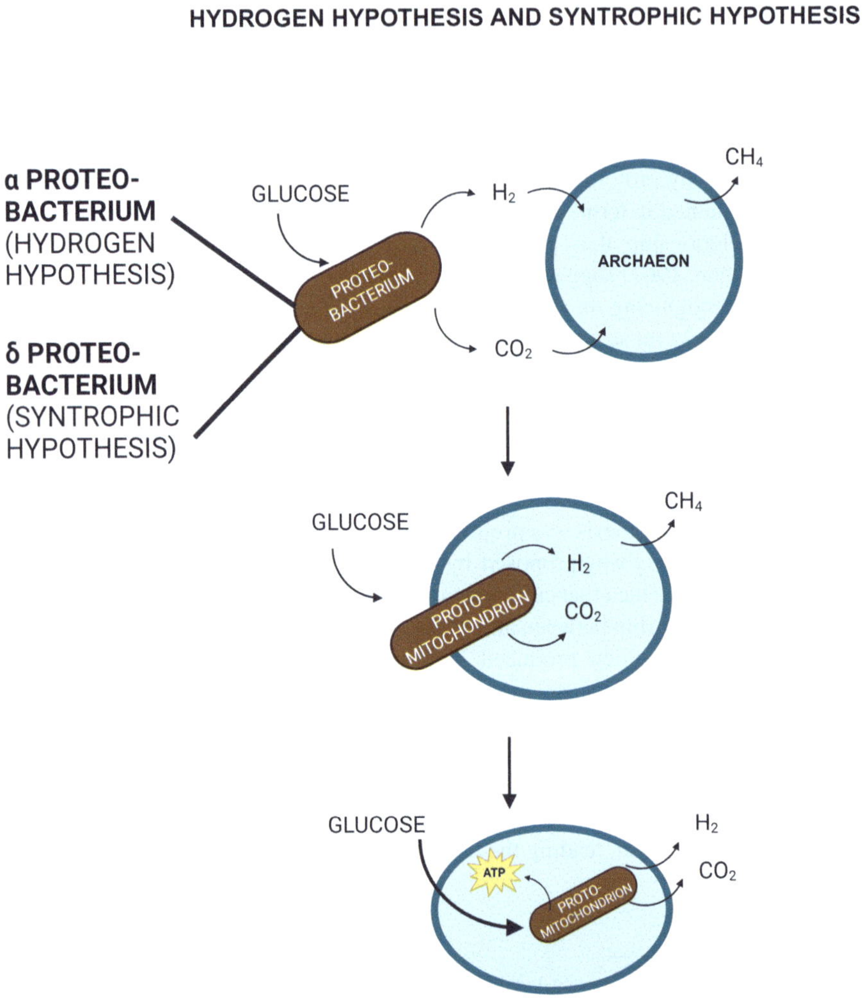

Fig. 4.2 The hydrogen and syntrophic hypotheses on mitochondrial origin. Both hypotheses suggest a symbiotic relationship between a hydrogen-producing proto-mitochondrion and a hydrogen-dependent archaeon. The proto-mitochondrion generates hydrogen (H_2) and carbon dioxide (CO_2) from glucose, which the host archaeon uses to produce methane (CH_4). At a certain point, a dramatic metabolic shift occurred, ceasing CH_4 production in favour of using O_2. The hydrogen hypothesis postulates that the proto-mitochondrion was an α-proteobacterium, whereas the syntrophic hypothesis posits that it was a δ-proteobacterium

Despite these similarities, the hydrogen and syntrophic hypotheses differ in two important respects. First, they propose different bacterial species for the proto-mitochondrion: Martin and Müller suggest an α-proteobacterium, while Moreira and López-García propose a δ-proteobacterium.[27] Second, Martin and Müller view the endosymbiosis as an obligate and strict dependence of the host on the endosymbiont, marking the first step in eukaryogenesis. In contrast, Moreira and López-García argue that endosymbiosis occurred in a host that already possessed some primitive eukaryotic features, such as a complex membrane system and a developed protoplasmic region.

Besides mitochondria, plastids also have an endosymbiotic origin (see Howe et al., 2008; Keeling, 2010). There is a general agreement that the ancestors of plastids were cyanobacteria that were engulfed by the proto-eukaryotic cell after the appearance of mitochondria (Archibald, 2009).[28] Plastids are monophyletic, meaning they share a common ancestor and originate from a single primary endosymbiotic event.[29] Their distribution across the eukaryotic tree of life results from secondary and tertiary endosymbiotic events, in which a heterotrophic eukaryotic cell engulfs a photosynthetic eukaryote. This process leads to the formation of chloroplasts in plants and green algae, rhodoplasts in red algae, and muroplasts in glaucophytes.

4.4.1 Endosymbiotic Models

Endosymbiotic models (Fig. 4.3a) place great emphasis on the endosymbiotic origin of mitochondria and the nucleus. There are different versions of endosymbiotic models depending on how endosymbiotic processes are explained. In general, endosymbiotic models do not address whether mitochondria appeared before or after the nucleus.

In 1991, Zillig proposed that eukaryogenesis was the result of "some sort of fusion event between an archaeal and one or several (possibly bacterial) ancestors" (Zillig, 1991, p. 545). The fusion model was initially proposed to explain the origin of mitochondria, but it leaves open the possibility that the nucleus also arose through fusion among prokaryotes. Zillig's hypothesis seeks to account for both the genetic

[27] In 2020, Moreira and Lopez-Garcia revised this hypothesis and proposed an updated model. They suggest that eukaryogenesis involved a tripartite symbiosis consisting of a sulphate-reducing δ-proteobacterium (the host), an Asgard-like archaeon that produced hydrogen (the precursor to the nucleus), and a sulphide-oxidising α-proteobacterium (the precursor to the mitochondrion) (López-García & Moreira, 2020).

[28] Margulis argued that chloroplast DNA derives from photosynthetic bacteria and represents "a multigenic hereditary system, which is not derived from the nucleus and is, in part, responsible for the biochemical properties of the organelle" (Sagan, 1967, p. 269).

[29] An exception to the primary endosymbiosis of plastids is the photosynthetic organelle of *Paulinella chromatophora*, an amoeboid protist. This organism acquired its photosynthetic organelles through a distinct primary endosymbiotic event.

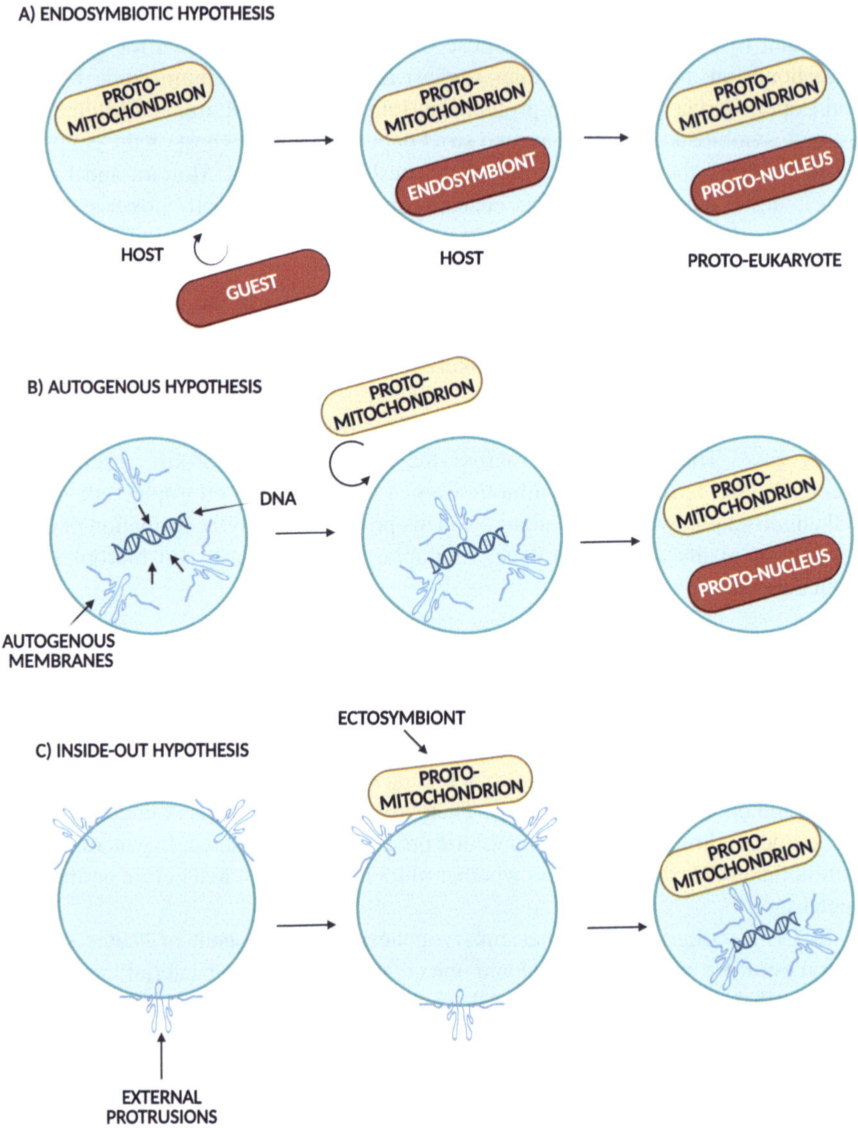

Fig. 4.3 Comparative overview of eukaryogenesis models. This figure illustrates three hypotheses regarding the origin of the nucleus and eukaryotic cells: (**a**) Endosymbiotic hypothesis: A host cell incorporates a guest cell, leading to a symbiotic relationship that results in the development of a proto-nucleus within the host. (**b**) Autogenous hypothesis: The nucleus forms internally from the host cell's own structures, independent of external entities, resulting in a proto-nucleus. (**c**) Inside-out hypothesis: External protrusions of the host cell enclose proto-mitochondria, which eventually leads to the formation of a proto-nucleus

chimerism of eukaryotes and the major biochemical innovations of the eukaryotic cell. The endosymbiont would have provided the archaeon (i.e. the proto-eukaryote) with the ability to synthesise fatty-acid ester lipids, whereas the archaeon developed isoprenyl-ether lipids.[30] According to Zillig, proto-eukaryotes possessed a large set of genes and enzymes sufficient for autonomous existence and the molecular components necessary for their accurate production (e.g. a DNA genome, RNA polymerases, and ribosome binding sites). Gradually, the massive transfer of genes from the endosymbiont(s) to the host, along with some changes in the ancestral communication system between the two symbionts, would have favoured the emergence of a unique, integrated organism.

In the same year, Sogin proposed an intriguing version of endosymbiosis. Building on Woese's (1987) thesis that bacteria, archaea, and eukaryotes share a common ancestor, the Progenote, Sogin (1991) argued that the Progenote was an RNA-based organism that split into two evolutionary lineages: the proto-eukaryotic and proto-bacterial lineages. The former developed a cytoskeleton, an RNA-based metabolism, a fragmented genome, and a refined RNA-based translation apparatus. The latter was characterised by the development of a protein-based metabolism, a DNA-based genome with loss of genome fragmentation, refined ribosomes, and a transcription system linked to translation. Sogin proposed that the appearance of the cytoskeleton allowed the proto-eukaryotic cell to develop phagocytosis, which in turn enabled it to engulf an archaeon. This endosymbiont was the ancestor of the nucleus and explains the genetic chimerism of eukaryotes, because "the archaebacterial genome contributed DNA and the majority of protein-coding regions found in contemporary eukaryotes, while the proto-eukaryotic lineage contributed coding information for the synthesis of both the translation apparatus and the cytoskeleton" (Sogin, 1991, p. 461).

The endosymbiotic origin of the nucleus was strongly supported by Lake et al. (1982) and Lake and Rivera (1994), who proposed the "endokaryotic hypothesis".[31] This account argued that the nucleus, "like the other eukaryotic organelles enclosed in double membranes (chloroplast and mitochondrion), has been derived through capture by an engulfing species" (Lake et al., 1982, p. 5951). According to Lake et al. (1982) and Lake and Rivera (1994), there was only a single engulfment process that gave rise to the nucleus, mitochondria, and chloroplasts. During this time, the guest (i.e. the proto-nucleus) took control of the host. This engulfment process involved a Gram-negative bacterium (the host) and an eocyte,[32] which was the guest.

In the same vein, Gupta and Bryan Golding (1996) suggested that the proto-eukaryotic cell arose through a unique endosymbiotic event. The host was a Gram-negative eubacterium, and the guest was an eocyte archaebacterium. The host eubacterium developed numerous membrane infolds that surrounded the guest,

[30] It is worth noting that the issue of lipid metabolism in the host, the symbiont and the resulting eukaryotic cell remains a complex and unresolved question.

[31] The endokaryotic hypothesis was proposed within the context of a comparative phylogenetic analysis of ribosomal small subunits in prokaryotes, archaea, and eukaryotes (Lake et al., 1982).

[32] Eocytes are a group of archaea discovered by Lake and colleagues (Lake et al., 1984).

which then transformed into the nucleus and the endoplasmic reticulum. As the membrane of the host enveloped the guest species, its own membrane (containing ether-linked lipids) became redundant and was lost. The formation of the nuclear envelope and endoplasmic reticulum was likely preceded or accompanied by the appearance of specific proteins (e.g. hsp70, hsp90) that facilitated communication between these organelles and the rest of the cell. The complete transformation of the two symbionts into a single organism required the transfer of genes from the host genome to the nucleus.

The endosymbiotic origin of the nucleus was also supported by Horiike et al. (2001, 2002, 2004), who proposed that the nucleus resulted from an endosymbiotic process between a eubacterium (the host) and an archaeon (the guest). They provided arguments based on the similarities of open reading frames (ORFs)[33] in eukaryotes and prokaryotes, suggesting that "ORFs in the functional categories related to the nucleus originated from archaea, while those related to the cytoplasm originated from eubacteria" (Horiike et al., 2004, p. 607).

A very specific and peculiar version of the endosymbiotic origin of eukaryotes was proposed by Margulis et al. (2000). They argued that the proto-eukaryotic cell resulted from a symbiotic process between thermoplasma-like archaebacteria and spirochaeta-like eubacteria. This association produced a proto-eukaryotic cell with a "Thiodendron stage", referring to a bacterial consortium (archaea + eubacteria) with a sulphur-based metabolism and motile capabilities (see Margulis et al., 2000, p. 6955). The authors suggested that the genetic material of the two organisms merged and was enclosed by a set of proteins produced by both organisms, forming the nuclear envelope. The appearance of the nucleus, along with the emergence of a single kinetosome[34] and its protein connector, led to the creation of an ancestral eukaryotic cell, which Margulis and colleagues named the "karyomastigont".[35]

An alternative version of the endosymbiotic origin of the nucleus was proposed by Takemura (2001) and Bell (2001, 2020). They argued that eukaryogenesis resulted from an endosymbiotic process between an archaeon and a bacterium, with the nucleus emerging from a viral infection of the archaeon. The host genome was thus destroyed and replaced by the viral genome.[36] In a similar vein, Forterre (2011) proposed that new eukaryotic proteins were the result of viral invasions of the proto-eukaryotic genome, introducing new genes and leading to the expression of new proteins. According to Forterre (2011), the first eukaryotic common ancestor

[33] ORFs stand for open reading frames. In molecular biology, ORFs are segments of DNA that extend from start to stop codons. In contemporary research, they are used to predict the presence of coding genes within a genomic sequence.

[34] The kinetosome is a protein complex located at the base of eukaryotic cilia or flagella, providing motility to eukaryotic cells.

[35] This term was coined by zoologist Konstantin Janicki to describe an organism with a flagellum-like appendage (the "mastigont") connected to the nucleus via a nuclear connect (see Margulis et al., 2000, pp. 6957–6958).

[36] According to Takemura (2001), the virus was an Orthopoxvirus, whereas Bell (2020) suggests that the virus belonged to the genus of Mimiviruses.

(FECA) arose from the engulfment of a Thaumarchaeon species by a PVC (Planctomycetes, Verrucomicrobia, and Chlamydiae) bacterium. In this scenario, the endosymbiont provided the host with all proteins and machinery of archaeal origin,[37] while the PVC bacterium supplied phospholipids, tubulin, and membrane coat proteins essential for the formation of the nucleus.

Some pitfalls can be identified in the endosymbiotic hypotheses. They often fail to clarify how major innovations of the nucleus, such as nuclear pores, the nuclear lamina, and nuclear import/export machineries, originated. For instance, if we hypothesise a single endosymbiotic event responsible for producing the nucleus, mitochondria, and chloroplasts (as proposed by Lake and Rivera (1994)), the relationships between these events remain unclear. Furthermore, the endosymbiotic origin of the nuclear envelope does not address how and why the endoplasmic reticulum is contiguous with the nucleus. Finally, most of these models do not adequately explain the emergence of other important eukaryotic features, such as the endomembranous system and secretory system, the cytoskeleton, and mitosis.

4.4.2 Autogenous Models

Cavalier-Smith (1975) first proposed a theory of eukaryogenesis based on the autogenous formation of the nucleus and endomembrane system (Fig. 4.3b).[38] He posited that the ancestor of eukaryotes was a unicellular, wall-free blue-green alga, which lost its cell wall and developed the capacity for phagocytosis. This development was a crucial step in eukaryogenesis because it provided "not only the selective pressure but also the physical mechanism (membrane budding and fusion) for cell compartmentation by intracellular membranes" (Cavalier-Smith, 1975, p. 463). Although phagocytosis reduced the surface area of the plasma membrane, it divided the intracellular space into phagosomal membranes, thus promoting intracellular compartmentation.

The development of phagocytosis is closely linked to the origin of the eukaryotic cytoskeleton and actin filaments. These structures provided the proto-eukaryote with a mechanism for cleavage (termed "cytosis"), which was foundational for "not only exocytosis, endocytosis, and cleavage but all forms of controlled membrane

[37] Forterre (2011) notes that the proteins provided by the thaumarchaeon included both informational and operational proteins (e.g. actin precursors and components of the vesicle trafficking system).

[38] Some authors (e.g. Thiergart et al., 2012; McInerney et al., 2014) characterise autogenous models as proposing that eukaryogenesis occurred through the stepwise mutation of a single ancestral lineage without external acquisition of genetic material. However, this characterisation is not entirely accurate, as autogenous models do acknowledge that mitochondria and chloroplasts have an endosymbiotic origin and that the eukaryotic genome acquired genes from bacteria and archaea. Thus, the primary distinction between endosymbiotic and autogenous models lies in their explanations of the origin of the eukaryotic nucleus.

budding and fusion in eukaryotes" (Cavalier-Smith, 1975, p. 464).[39] Furthermore, the appearance of the eukaryotic cytoskeleton introduced mechanisms for mitosis, which evolved before the nucleus and ensured the "segregation of the two sister centrosome/DNA complexes at division" (Cavalier-Smith, 1988, p. 75). Mitosis is believed to have begun with the attachment of centrosome microtubules to the cell surface, leading to the formation of a proto-cilium.[40]

According to Cavalier-Smith (1975, 1987, 1988, 2010), the nuclear and endoplasmic reticulum membranes share a common origin in phagocytosis, which created intracellular compartments with "both DNA and ribosomes attached to their cytoplasmic faces" (Cavalier-Smith, 1988, p. 74). The nuclear envelope is thought to have evolved from a transformation of endomembranes containing DNA molecules. This transformation involved the formation of the nuclear lamina as an initial step, facilitating the attachment of endomembranes to chromatin.[41] Additionally, the emergence of the nucleus allowed for the presence of multiple replicons[42] that replicate simultaneously, thereby reducing selective constraints on genome size by increasing the number, but not the length, of replicons (see Cavalier-Smith, 1988, p. 77).

Phagocytosis and the endomembrane system developed before the proto-eukaryotic cell acquired cellular endosymbionts, which later became mitochondria and chloroplasts (Cavalier-Smith, 1987, 1988). Mitochondria originated from an aerobic proteobacterium that was engulfed, either as prey or as a parasite, and became an endosymbiont providing the host with the metabolic capability to thrive in an oxygen-rich environment. Some eukaryotes subsequently acquired new endosymbionts, specifically cyanobacteria, which were later transformed into plastids.

Like Cavalier-Smith, Taylor (1976) supported the autogenous origin of eukaryotes. He proposed that eukaryotes originated from a cyanophyte-like photosynthetic ancestor, termed the "Uralga". This organism progressively developed differentiated membrane systems, creating double-membrane metabolic compartments. Taylor (1976) concurred with Cavalier-Smith that the evolution of endomembranes necessitated the development of endo- and exocytic properties. In this scenario, mitochondria would have originated autogenously, while plastids would have arisen through endosymbiosis.[43]

---

[39] Cavalier-Smith argued that endocytosis and exocytosis evolved together, as they are mutually dependent (Cavalier-Smith, 1975, p. 463).

[40] This idea is based on the observation that some unicellular eukaryotes (e.g. amoebae) have a nucleus attached to a single basal body.

[41] The nuclear lamina provides three selective advantages: it protects DNA from shearing damage caused by cytoplasmic contractility; it facilitates the attachment of endomembranes to segregating DNA; and it increases metabolic efficiency by concentrating the protein machinery involved in DNA replication, transcription, and RNA processing within a confined volume.

[42] Replicons are a region of DNA that replicates independently from a single origin of replication.

[43] The main steps of eukaryogenesis are detailed as follows: "first, a dispersed genome and multi-purpose lamellae; secondly, formation of a larger nucleoid; thirdly, formation of a nuclear membrane from the innermost thylakoids; differentiation of some membranes for specialized functions [...]; an engulfment phase, dependant on changed membrane properties and the association of

Autogenous models are valuable for providing a critical perspective on endosymbiotic hypotheses and emphasising the autogenous origin of the nucleus and endomembranous system. However, a potential limitation of these models is their tendency to downplay the significance of mitochondria in eukaryogenesis. They typically position the origin of mitochondria as a later event, occurring after the development of phagocytic abilities and mitosis. Cavalier-Smith (1987, 1988) further expanded on this by proposing the archezoa hypothesis, which suggests that primitively amitochondriate proto-eukaryotes are living relics of the transition from prokaryotes to eukaryotes. Discoveries of hydrogenosomes (Lindmark & Müller, 1973) and mitosomes (Tovar et al., 1999) have revealed that these organelles, involved in iron-sulphur cluster biosynthesis, have an endosymbiotic origin. This indicates that the proto-eukaryotic cell may have acquired endosymbionts early in eukaryotic evolution, which could have played a crucial role in the autogenous formation of the nucleus and endomembranous system (Martin & Müller, 1998; Lane, 2011, 2014).

4.4.3 The Inside-out Model

Both endosymbiotic and autogenous models view mitochondria (and, in the case of endosymbiotic hypotheses, the nucleus) as resulting from the progressive transformation of endosymbionts into organelles. They propose that symbionts originated externally and entered the host through the development of an endosymbiotic relationship. In contrast to these "outside-in" hypotheses, Baum and Baum (2014) proposed an "inside-out" model of eukaryogenesis (Fig. 4.3c).[44]

This theoretical account suggests that the fusion between two prokaryotes occurred gradually, without a single phagocytic event. It posits that proto-mitochondria were ectosymbionts[45] of an ancestral prokaryotic cell, which produced extracellular protrusions that engulfed these ectosymbionts, leading to their transformation into organelles. According to the inside-out model, the outer nuclear membrane, plasma membrane, and cytoplasm originated from these extracellular protrusions, while the endoplasmic reticulum formed from the spaces between them. The final step in eukaryogenesis was the formation of a continuous plasma membrane that isolated the endoplasmic reticulum from the exterior. The

microfibrils and microtubules, during which mitochondrial and chloroplasts regions become isolated; sixth, further differentiation of the primary cell matrix and membranes for greater endocytosis and exocytosis; seventh, formation of 9 + 2 microtubule-organizing centres from simpler primordia; and eighth, flagella formation" (Taylor, 1976, p. 383).

[44] A somewhat similar approach was proposed earlier by de Roos (2006), who suggested that a double-membrane proto-eukaryote secreted membranous extracellular vesicles that fused to form the plasma membrane (see Baum & Baum, 2014, p. 4).

[45] Ectosymbionts (or epibiotic organisms) are those that establish a symbiotic relationship with their host by living on its surface, as opposed to endosymbionts, which live within the host.

proto-eukaryotic cell would have acquired an early proto-mitotic cycle, enabling it to reproduce as a whole.[46]

In the inside-out model, the host was an eocyte ancestor,[47] and mitochondria were epibiotic α-proteobacteria. The eocyte cell developed protrusions[48] that facilitated material exchange with the proto-mitochondria.[49] Over time, the area of contact between the symbionts increased, leading to the enlargement of these protrusions and the eventual loss of surface-layer[50] material from them. Stable protrusions enhanced the cell's surface area-to-volume ratio, thereby improving material exchange with the environment. The enlargement of the protrusions and the thinning or loss of cell wall material resulted in the formation of extracellular blebs. These blebs produced vesicles that fused together, leading to the formation of the nuclear envelope and the endoplasmic reticulum.

This process facilitated the curvature of the extracellular blebs, which was stabilised by certain proteins (e.g. Coat Protein Complex II) and led to the formation of nuclear pores.[51] The recruitment of additional proteins (e.g. homologues of structural nucleoporins) to these nuclear pore complexes allowed the proto-eukaryotic cell to stabilise the protrusions and promote the development of an outer ring of nucleoporins. This ring supported a second area of positive curvature outside the cell wall, resulting in the formation of the nuclear envelope.

The nuclear membrane enhanced the regulation of lipid and aqueous material flow between the nuclear and cytoplasmic compartments. It also regulated the transport of mRNA and ribosomes, creating two distinct domains for protein translation: nuclear and cytoplasmic. The controlled transport of mRNA and proteins between these compartments enabled the eukaryotic cell to segregate core metabolic processes in the cytoplasm from DNA replication, transcription, and ribosome assembly in the nucleoplasm.

The origin of the endoplasmic reticulum and its continuity with the nuclear envelope is readily explained: the progressive growth of extracellular protrusions created internal pressures that formed "a continuous network of inter-bleb cripts,

[46] Baum and Baum argue that some proteins (such as ESCRTIII, LINC complexes, and dynamin) would have facilitated the scission of cytoplasmic bridges between cells, the transmission of nuclear organisation, and mitochondrial segregation after cell division (see Baum & Baum, 2014, p. 9).

[47] Eocytes were likely characterised by a single bounding membrane and a glycoprotein-rich cell wall (Baum & Baum, 2014).

[48] Protrusions might have been enabled by specific molecular machineries (e.g. proteins homologous to COPII proteins) that drove the budding of endomembrane vesicles (Baum & Baum, 2014).

[49] The nature of material exchange between these symbiotic partners is still debated, but Baum and Baum (2014) propose that it could have included hydrogen, ATP, sulphur, hydrogen sulphide, and organic acids.

[50] The surface-layer (S-layer) is a component of the cell envelope found in almost all archaea and many bacteria. It consists of a monomolecular layer of one (or two) identical proteins or glycoproteins.

[51] Proteins originally involved in stabilising positive membrane curvature in the nuclear pore were later repurposed for vesicle formation.

homologous to the lumen of the nuclear envelope and the ER of modern eukaryotes" (Baum & Baum, 2014, p. 6). The rough endoplasmic reticulum would have arisen from the relocation of proteins synthesised by ribosomes into the perinuclear lumen and the proximal endoplasmic reticulum.[52] Subsequently, the continuity between the endoplasmic reticulum and the Golgi apparatus was facilitated by the cytoskeleton and the thinning or loss of residual cell wall material (Baum & Baum, 2014, p. 10).

According to the authors, phagocytosis evolved after the acquisition of mitochondria. This required the transfer of genes responsible for synthesising lipids present in mitochondria to the proto-eukaryotic cell. The development of phagocytosis involved physico-chemical modifications of the plasma membrane, making it more dynamic and endowing it with reversible deformation capabilities.

In contrast to outside-in models (endosymbiotic and autogenous hypotheses), the inside-out model provides a novel and heuristic framework for explaining eukaryogenesis. Baum and Baum (2014) highlight three advantages of this model. Firstly, it offers a parsimonious explanation for specific eukaryotic features without relying on ad hoc assumptions. Secondly, it accounts for unusual eukaryotic characteristics, such as the extensive continuity between the nucleus and the endoplasmic reticulum. Lastly, it aligns with phylogenetic analyses suggesting that the proto-eukaryotic cell was an eocyte and that secretory pathways and exocytosis evolved prior to endocytosis and phagocytosis.[53]

4.5 Biological Individuality and Autonomy in Eukaryogenesis

The previous sections have elucidated why eukaryogenesis represents a major evolutionary transition and have examined the key controversial issues and hypotheses regarding the origin of eukaryotic cells. Beyond the biological questions, several important philosophical and conceptual issues emerge. This section integrates philosophical arguments about biological individuality (discussed in Chap. 1) and autonomy (discussed in Chap. 2) with the biological issues underlying eukaryogenesis.

Biological individuality is central to the concept of eukaryogenesis, as it involves the transformation of one form of individual (the prokaryotic cell) into a fundamentally new one. On the one hand, there is significant continuity between these two forms, evidenced by the retention—albeit with modifications—of prokaryotic structures and functions, and the presence of numerous genes of prokaryotic origin in

[52] This process may have been facilitated by SRP (signal recognition particle) complexes of archaeal origin, which were concentrated on the membranes of protrusions close to the original cell body (Baum & Baum, 2014, see, p. 10).

[53] These analyses are based on the phylogenetic examination of the Ras GTPase superfamily (Jékely, 2003).

eukaryotes. On the other hand, the emergence of a radically new biological architecture and new functional properties signifies a pivotal shift in the evolution of life. This break in continuity compels us to examine how a physiological and evolutionary individual (see Sect. 1.4) emerged during eukaryogenesis. Specifically, what conditions allowed the proto-eukaryotic cell to evolve into both a physiological unit and a unit of selection?

As discussed in Sect. 1.4.2, various criteria have been proposed to define physiological individuality. These include the metabolic unit (Jonas, 1966; Dupré & O'Malley, 2009; Landecker, 2017), the presence of an immune system (Pradeu, 2010, 2016), a low level of conflict coupled with a high degree of cooperation among the parts (Gardner & Grafen, 2009; Queller & Strassmann, 2009), internal functional differentiation (Folse & Roughgarden, 2010; Arnellos et al., 2014), and functional interdependence among different levels of organisation (Zylstra, 1992; Korn, 2002). These criteria are evident in the three primary models of eukaryogenesis, illustrating the close relationship between the emergence of a new biological organisation and the establishment of physiological unity.

The origin of mitochondria and plastids through engulfment is a common tenet in all three models. This evolutionary step is both crucial and problematic in the development of a physiological unit. On the one hand, the endosymbiont and host likely formed a cooperative metabolic relationship (see Martin & Müller, 1998; Moreira & Lopez-Garcia, 1998), thereby establishing a metabolic unit. On the other hand, the formation of a long-term, stable relationship between these two organisms was fraught with challenges. The engulfment of a prokaryote by another involved significant conflicts between their genetic elements, metabolic pathways, and immune systems (Radzvilavicius & Blackstone, 2015). Thus, the emergence of physiological individuality required strategies to resolve or mitigate these conflicts and enhance cooperation between the organisms (see Michod & Nedelcu, 2003; Queller & Strassmann, 2009). Effective strategies for overcoming biological conflicts included the extensive transfer of genes from the endosymbiont to the host, and the introduction of new protein complexes,[54] regulatory pathways, and signalling mechanisms[55] that facilitated communication and functional interdependence between the two organisms (Radzvilavicius & Blackstone, 2015).

The appearance of the endomembrane system, which is interpreted differently across the three eukaryogenesis theories, helps clarify some conditions for the emergence of a physiological individual. Conceptually, the internal division of the intracellular space into endomembranes presents challenges similar to those encountered with the origin of mitochondria and plastids. On the one hand, the development of new organelles (e.g. nucleus, endoplasmic reticulum, Golgi apparatus) facilitated internal functional differentiation, a crucial aspect of physiological individuality (Folse & Roughgarden, 2010; Arnellos et al., 2014). On the other hand,

[54] For instance, the mitochondrial protein import machinery is a novel protein complex that supported the integration of proto-mitochondria into the protoeukaryotic cell.

[55] The calcium signalling pathway used to depolarise proto-mitochondria could serve as a representative example (Radzvilavicius & Blackstone, 2015).

the endomembrane system introduced new levels of intracellular organisation, which could complicate the coordination of the proto-eukaryotic cell's internal components. As I will discuss in Chap. 6, a key step in the evolution of physiological individuality was the development of a sophisticated system for secretory particle and protein transport, which enabled effective communication among the various organelles within the cell.

Another significant aspect highlighted by the three models of eukaryogenesis is the role of the eukaryotic cytoskeleton in achieving physiological integration. The cytoskeleton promoted cooperation and functional interdependence among cellular components. The emergence of actin filaments, microtubules, and intermediate filaments allowed the eukaryotic cell to regulate intracellular transport, maintain its shape, and control cellular movement. Additionally, as suggested in Chap. 7, the integration of endosymbionts into the host cell and their transformation into organelles were facilitated by the eukaryotic cytoskeleton. The motility of the endosymbionts was progressively and fully managed by the host, thereby converting their biological capabilities, particularly metabolic functions, into assets beneficial for the host cell.

Sect. 1.4.1 has outlined several criteria for defining evolutionary individuality, including the following: an entity exhibiting phenotypic variation, differential fitness, and heritable fitness (Lewontin, 1970); a replicator (Dawkins, 1976); a replicator plus an interactor (Hull, 1980); an entity with a developmental system (Griffiths & Gray, 1994); a developmental system combined with a reproductive system (Griesemer, 2000, 2016); a reproductive system that transmits variation to offspring (Godfrey-Smith, 2013); and an entity displaying genetic uniqueness, homogeneity, and autonomy (Santelices, 1999). These criteria are also relevant to the models of eukaryogenesis and help to elucidate how the proto-eukaryotic cell evolved into an evolutionary individual.

For the proto-eukaryotic cell to qualify as a replicator, it needed the capacity to reproduce as a whole and to transmit its genetic and phenotypic traits—including its organelles—to its offspring through vertical transmission. As Griesemer (2000, 2016) has highlighted, this necessitated the development of a reproductive system integrated with the cell's developmental processes. In this context, the transition from prokaryotic binary fission to eukaryotic mitosis marked a significant evolutionary milestone. Mitosis enabled the proto-eukaryotic cell to reproduce as a whole, reconstituting its architecture and ensuring the inheritance of its organelles by daughter cells. The evolutionary origin of mitosis remains complex and is not fully understood: since it relies heavily on cytoskeletal proteins and the division of chromosomes within a cell nucleus, there is ongoing debate about whether it emerged earlier or later in eukaryogenesis.

For example, both endosymbiotic (Sogin, 1991) and autogenous models (Cavalier-Smith, 1987, 1988) propose that mitosis was an early event in eukaryogenesis, directly linked to the appearance of the eukaryotic cytoskeleton. In contrast, the inside-out model argues that the development of the eukaryotic cytoskeleton and mitosis occurred later, following the formation of early membrane protrusions (Baum & Baum, 2014).

In addition to the origin of mitosis, another crucial step in the evolution of evolutionary individuality during eukaryogenesis was the capacity for vertical transmission of endosymbionts, such as proto-mitochondria and proto-chloroplasts. In biological systems, the engulfment of a prokaryote by another typically results in a transient infection (Corsaro & Venditti, 2006), which is not passed from parent to offspring. Long-term endosymbiosis, involving a stable relationship between two prokaryotes, is exceedingly rare[56] and has been documented only between specific classes of proteobacteria (von Dohlen, 2001). In such cases, the endosymbiont not only loses many of its genes and becomes unable to survive independently but also undergoes vertical transmission, ensuring its inheritance through generations (see Chap. 5 for further details).

Eukaryogenesis is intricately linked to both biological individuality and autonomy. As discussed in Chap. 2, biological autonomy is divided into two key aspects: constitutive autonomy and interactive autonomy. Constitutive autonomy refers to a system's ability to self-maintain and self-sustain within meta-stable conditions, while interactive autonomy denotes an entity's capacity to act as an agent—interacting with and modifying its environment (Moreno & Mossio, 2015).

Interestingly, current theories of eukaryogenesis do not explicitly address how an autonomous biological organisation emerges during major evolutionary transitions. The dominant explanatory strategies in these theories typically focus on the sequence of structural and functional changes that led to the formation of the modern eukaryotic cell. These theories rely on phylogenetic evidence and hypotheses about the probability of various evolutionary events rather than explicitly connecting these transitions to concepts of biological autonomy. Similarly, Maynard Smith and Szathmáry (1995), while exploring transitions in individuality and complexity, do not directly link these changes to the development of biological autonomy.

Despite this, the models of eukaryogenesis implicitly acknowledge the development of metabolic autonomous capacities and the ability of the eukaryotic cell to interact with and transform its environment. The concept of biological autonomy, therefore, underpins these theories, even if it is not explicitly articulated.

To understand how eukaryogenesis facilitated the development of constitutive and interactive autonomy, we can use an organisational approach that emphasises the role of constraints and their interdependence (or closure) in fostering both types of autonomy (Moreno & Mossio, 2015). Applying this framework to eukaryogenesis, we must consider which *structural constraints* enabled the emergence of *constitutive* and *interactive autonomy*.

The engulfment of proto-mitochondria represents a crucial structural constraint with significant implications for both the endosymbiont and the host, profoundly affecting the self-maintenance of the proto-eukaryote. This event established a common boundary, the membrane of the host, which necessitated the sharing of metabolic pathways and fostered an obligate symbiosis between the host and the

56 In contrast, the endosymbiosis of prokaryotes within eukaryotic multicellular organisms (e.g. some insect species) is quite common.

endosymbiont. As a result, selective pressures were imposed on the endosymbiont, leading to the progressive transfer of genes from the endosymbiont to the host. This gene transfer diminished the endosymbiont's autonomy and facilitated its transformation into an organelle (see Chap. 5).

Similarly, the emergence of the endomembrane system introduced new constraints on the biological processes essential for the self-maintenance of proto-eukaryotes. For instance, the development of the nucleus enabled the separation of gene transcription, which occurred inside the nucleus, from translation, which took place outside in the ribosomes. This decoupling allowed for more precise regulation of gene expression. The constraints that led to the formation of the endomembrane system are interpreted differently by the three main theories of eukaryogenesis. Endosymbiotic theories focus on the invagination process resulting from endosymbiotic events, suggesting that the endomembrane system originated from membrane-bound structures associated with endosymbionts. Autogenous theories, in contrast, emphasise the internal pressures exerted by the plasma membrane of the proto-eukaryotic cell, proposing that these pressures led to the formation of internal membranes and organelles. The inside-out model highlights the internal pressures created by external protrusions on the cell's interior, arguing that these protrusions contributed to the development of the endomembrane system and the formation of organelles. Each model offers a distinct perspective on how these constraints contributed to the evolution of new forms of biological organisation and autonomy in eukaryotes.

The emergence of interactive autonomy in eukaryogenesis is intricately linked to the constraints imposed by the eukaryotic cytoskeleton on the proto-eukaryotic cell. These constraints led to significant advancements in cellular movement, contractility, and vesicular transport. The development of the eukaryotic cytoskeleton was pivotal for the transformation of endosymbionts into organelles. It allowed the host cell to control and utilise the motility of the endosymbionts, effectively integrating their capabilities into the host's metabolic and cellular functions (Militello, 2019).

One of the most profound consequences of the emergence of the eukaryotic cytoskeleton was the origin of mitosis. The evolution from prokaryotic binary fission to eukaryotic mitosis was likely driven by the need to reorganise the intracellular space functionally and to ensure the accurate transmission of this organisation to offspring. Prokaryotic cytoskeletal proteins were insufficient for this task, necessitating the development of a more complex mitotic mechanism. As discussed in Chap. 8, this transition was crucial for the proto-eukaryotic cell to pass on its material organisation to progeny, thereby exhibiting cross-generational functions and autonomy at the species level (see Chap. 2).

4.6 Concluding Remarks

This chapter has explored the primary biological and philosophical issues surrounding eukaryogenesis. The transition from prokaryotic to eukaryotic cells is considered a major evolutionary transition in the sense described by Maynard-Smith and

Szathmáry (1995), as it involves an increase in structural and functional complexity and the emergence of a new form of individuality. Unlike other major evolutionary transitions, eukaryogenesis is unique in several ways: it occurred only once in the history of life and involved a process of endosymbiosis between prokaryotes, which significantly contributed to the development of a new kind of individuality.

Despite extensive study, eukaryogenesis remains mysterious in many aspects. The phylogenetic origin of eukaryotes is not entirely clear, and the eukaryotic genome is characterised by a mix of genes from both bacteria and archaea. This genetic chimerism suggests that a symbiotic relationship between prokaryotes occurred, involving a massive transfer of genes from one organism to another, followed by genetic variation. Another important feature of eukaryotes is the increase in genome size and the emergence of new biological functions, which is difficult to explain solely by an increase in cellular size. Indeed, there are reports of giant bacteria with relatively small genomes, indicating that other factors must have played a role in the evolution of eukaryotic complexity.

The sequence of events leading to the formation of eukaryotic cells is not yet fully understood, with three main theoretical models proposed: the endosymbiotic, autogenous, and inside-out hypotheses. All three agree that mitochondria and chloroplasts were acquired through a process of engulfment, which established a mutualistic symbiosis between the host (the proto-eukaryotic cell) and the guest (the proto-mitochondrion). However, they disagree on the mechanisms that facilitated this engulfment and on whether the acquisition of mitochondria was an early event in eukaryogenesis preceding the formation of the nucleus or occurred later. Additionally, these models differ on the mechanisms responsible for the origin of the nucleus and endomembrane system, suggesting either endosymbiotic or autogenous processes.

Eukaryogenesis offers insights into the origins of physiological and evolutionary individuality and indicates that the development of new structures (such as mitochondria, the endomembrane system, and the cytoskeleton) was accompanied by potential conflicts and challenges. Similarly, the emergence of eukaryotic organelles illuminates the concepts of constitutive and interactive autonomy: the transition from prokaryotic to eukaryotic cells involved the loss of autonomy of the endosymbionts and a significant reorganisation of the host's functional structure. This enabled proto-eukaryotes to develop new methods of self-maintenance and interaction with their environment, thereby fostering new forms of biological autonomy.

The emergence of a new biological individual during eukaryogenesis raises an intriguing question: how is functional integration achieved in such a way as to produce a physiological and autonomous unit with open-ended evolutionary capacities? This major evolutionary transition represents a dramatic shift from one type of functionally integrated organisation to another. The three leading hypotheses on eukaryogenesis propose distinct major changes that contributed to the establishment of a new functionally integrated organisation. These include the endosymbiotic processes leading to the development of mitochondria and the nucleus (as suggested by endosymbiotic theories), the evolution of phagocytosis and the eukaryotic cytoskeleton (as put forward by autogenous models), and the formation of extracellular protrusions that engulfed ectosymbionts (as described by the inside-out model). These changes exerted selective pressures that led to a cascade of events, ultimately resulting in a shift in functional integration.

Despite their differences, these three models indicate that the origin of organelles is linked with changes in how gene expression, metabolism, intra- and inter-cellular communication, and the life cycle are integrated. In Chaps. 5 through 8, I will argue that the emergence of each new eukaryotic structure (i.e. mitochondria, the endomembrane system, and the cytoskeleton) provides fundamental insights into the multifaceted functional integration of eukaryotes. In other words, examining these three structures can help explain the transition from one type of functional integration to another and provide an answer to the overarching question: what is functional integration, and which biological dimensions define it?

Interpreting eukaryogenesis as a transition in functional integration raises an obvious question: despite the differences, is it possible to identify a common core of functional integration that remains unchanged during the transition from prokaryotic to eukaryotic cells? In Chap. 5, I will explore the fundamental level of cellular functional integration shared by both prokaryotic and eukaryotic cells. This conceptual step is crucial for understanding how eukaryotic innovations emerged and how a new type of functional integration developed within the pre-existing prokaryotic framework of functional integration.

References

Archibald, J. M. (2009). The puzzle of plastid evolution. *Current Biology, 19*(2), R81–R88. https://doi.org/10.1016/j.cub.2008.11.067

Arnellos, A., Moreno, A., & Ruiz-Mirazo, K. (2014). Organizational requirements for multicellular autonomy: Insights from a comparative case study. *Biology & Philosophy, 29*(6), 851–884. https://doi.org/10.1007/s10539-013-9387-x

Baum, D. A. (2015). A comparison of autogenous theories for the origin of eukaryotic cells. *American Journal of Botany, 102*(12), 1954–1965. https://doi.org/10.3732/ajb.1500196

Baum, D. A., & Baum, B. (2014). An inside-out origin for the eukaryotic cell. *BMC Biology, 12*, 76. https://doi.org/10.1186/s12915-014-0076-2

Bell, P. J. L. (2001). Viral Eukaryogenesis: Was the ancestor of the nucleus a complex DNA virus? *Journal of Molecular Evolution, 53*(3), 251–256. https://doi.org/10.1007/s002390010215

Bell, P. J. L. (2020). Evidence supporting a viral origin of the eukaryotic nucleus. *Virus Research, 289*, 198168. https://doi.org/10.1016/j.virusres.2020.198168

Brocks, J. J., Logan, G. A., Buick, R., & Summons, R. E. (1999). Archean molecular fossils and the early rise of eukaryotes. *Science, 285*(5430), 1033–1036. https://doi.org/10.1126/science.285.5430.1033

Brueckner, J., & Martin, W. F. (2020). Bacterial genes outnumber archaeal genes in eukaryotic genomes. *Genome Biology and Evolution, 12*(4), 282–292. https://doi.org/10.1093/gbe/evaa047

Cavalier-Smith, T. (1975). The origin of nuclei and of eukaryotic cells. *Nature, 256*(5517), 463–468. https://doi.org/10.1038/256463a0

Cavalier-Smith, T. (1987). The origin of eukaryotic and archaebacterial cells. *Annals of the New York Academy of Sciences, 503*, 17–54. https://doi.org/10.1111/j.1749-6632.1987.tb40596.x

Cavalier-Smith, T. (1988). Origin of the cell nucleus. *BioEssays: News and Reviews in Molecular, Cellular and Developmental Biology, 9*(2–3), 72–78. https://doi.org/10.1002/bies.950090209

Cavalier-Smith, T. (2010). Origin of the cell nucleus, mitosis and sex: Roles of intracellular coevolution. *Biology Direct, 5*, 7. https://doi.org/10.1186/1745-6150-5-7

Corsaro, D., & Venditti, D. (2006). Bacterial endosymbionts in prokaryotes. In J. M. Shively (Ed.), *Complex intracellular structures in prokaryotes* (pp. 359–371). Springer. https://doi.org/10.1007/7171_035

Dawkins, R. (1976). *The selfish gene*. Oxford University Press.

von Dohlen, C. D., Kohler, S., Alsop, S. T., & McManus, W. R. (2001). Mealybug Beta-Proteobacterial endosymbionts contain gamma-Proteobacterial symbionts. *Nature, 412*(6845), 433–436. https://doi.org/10.1038/35086563

Dupré, J., & O'Malley, M. (2009). Varieties of living things : Life at the intersection of lineage and metabolism. *Philosophy, Theory, and Practice in Biology, 1 (3): 1*, e003. https://doi.org/10.3998/ptb.6959004.0001.003

Eme, L., Sharpe, S. C., Brown, M. W., & Roger, A. J. (2014). On the age of eukaryotes: Evaluating evidence from fossils and molecular clocks. *Cold Spring Harbor Perspectives in Biology, 6*(8), a016139. https://doi.org/10.1101/cshperspect.a016139

Ferla, M. P., Cameron Thrash, J., Giovannoni, S. J., & Patrick, W. M. (2013). New rRNA gene-based phylogenies of the Alphaproteobacteria provide perspective on major groups. *Mitochondrial Ancestry and Phylogenetic Instability'. PloS One, 8*(12). https://doi.org/10.1371/journal.pone.0083383

Folse, H. J., & Roughgarden, J. (2010). What is an individual organism? A multilevel selection perspective. *The Quarterly Review of Biology, 85*(4), 447–472. https://doi.org/10.1086/656905

Forterre, P. (2011). A new fusion hypothesis for the origin of Eukarya: Better than previous ones, but probably also wrong. *Research in Microbiology, 162*(1), 77–91. https://doi.org/10.1016/j.resmic.2010.10.005

Gabaldón, T. (2021). Origin and early evolution of the eukaryotic cell. *Annual Review of Microbiology, 75*, 631–647. https://doi.org/10.1146/annurev-micro-090817-062213

Gardner, A., & Grafen, A. (2009). Capturing the superorganism: A formal theory of group adaptation. *Journal of Evolutionary Biology, 22*(4), 659–671. https://doi.org/10.1111/j.1420-9101.2008.01681.x

Godfrey-Smith, P. (2013). Darwinian individuals. In F. Bouchard & P. Huneman (Eds.), *From groups to individuals: Evolution and emerging individuality* (pp. 17–36). MIT Press.

Golding, G. B., & Gupta, R. S. (1995). Protein-based phylogenies support a chimeric origin for the eukaryotic genome. *Molecular Biology and Evolution, 12*(1), 1–6. https://doi.org/10.1093/oxfordjournals.molbev.a040178

Griesemer, J. (2000). Development, culture, and the units of inheritance. *Philosophy of Science, 67*(S3), S348–S368. https://doi.org/10.1086/392831

Griesemer, J. (2016). Reproduction in complex life cycles: Toward a developmental reaction norms perspective. *Philosophy of Science, 83*(5), 803–815. https://doi.org/10.1086/687865

Griffiths, P. E., & Gray, R. D. (1994). Developmental systems and evolutionary explanation. *Journal of Philosophy, 91*(6), 277–304. https://doi.org/10.2307/2940982

Gupta, R. S., & Bryan Golding, G. (1996). The origin of the eukaryotic cell. *Trends in Biochemical Sciences, 21*(5), 166–171.

Han, T. M., & Runnegar, B. (1992). Megascopic eukaryotic algae from the 2.1-billion-year-old Negaunee iron-formation, Michigan. *Science, 257*(5067), 232–235. https://doi.org/10.1126/science.1631544

Horiike, T., Kazuo Hamada, S., & Kanaya, and T. Shinozawa. (2001). Origin of eukaryotic cell nuclei by Symbiosis of archaea in bacteria is revealed by homology-hit analysis. *Nature Cell Biology, 3*(2), 210–214. https://doi.org/10.1038/35055129

Horiike, T., Hamada, K., Miyata, D., & Shinozawa, T. (2004). The origin of eukaryotes is suggested as the Symbiosis of Pyrococcus into γ-Proteobacteria by phylogenetic tree based on gene content. *Journal of Molecular Evolution, 59*(5), 606–619. https://doi.org/10.1007/s00239-004-2652-5

Horiike, T., Hamada, K., & Shinozawa, T. (2002). Origin of eukaryotic cell nuclei by Symbiosis of archaea in bacteria supported by the newly clarified origin of functional genes. *Genes & Genetic Systems, 77*(5), 369–376. https://doi.org/10.1266/ggs.77.369

Howe, C. J., Barbrook, A. C., Ellen, R., Nisbet, R., Lockhart, P. J., & Larkum, A. W. D. (2008). The origin of plastids. *Philosophical Transactions of the Royal Society B: Biological Sciences, 363*(1504), 2675–2685. https://doi.org/10.1098/rstb.2008.0050

Hull, D. L. (1980). Individuality and selection. *Annual Review of Ecology, Evolution, and Systematics, 11*, 311–332. https://doi.org/10.1146/annurev.es.11.110180.001523

Javaux, E. J. (2007). The early eukaryotic fossil record. *Advances in Experimental Medicine and Biology, 607*, 1–19. https://doi.org/10.1007/978-0-387-74021-8_1

Jékely, G. (2003). Small GTPases and the evolution of the eukaryotic cell. *BioEssays: News and Reviews in Molecular, Cellular and Developmental Biology, 25*(11), 1129–1138. https://doi.org/10.1002/bies.10353

Jonas, H. (1966). *The phenomenon of life: Toward a philosophical biology*. Harper & Row.

Katscher, F. (2004). The history of the terms prokaryotes and eukaryotes. *Protist, 155*(2), 257–263. https://doi.org/10.1078/143446104774199637

Keeling, P. J. (2010). The endosymbiotic origin, diversification and fate of plastids. *Philosophical Transactions of the Royal Society B: Biological Sciences, 365*(1541), 729–748. https://doi.org/10.1098/rstb.2009.0103

Koonin, E. V. (2010). The origin and early evolution of eukaryotes in the light of Phylogenomics. *Genome Biology, 11*(5), 209. https://doi.org/10.1186/gb-2010-11-5-209

Korn, R. W. (2002). Biological hierarchies, their birth, death and evolution by natural selection. *Biology and Philosophy, 17*(2), 199–221. https://doi.org/10.1023/A:1015271707591

Kowallik, K. V., & Martin, W. F. (2021). The origin of Symbiogenesis: An annotated English translation of Mereschkowsky's 1910 paper on the theory of two plasma lineages. *Biosystems, 199*, 104281. https://doi.org/10.1016/j.biosystems.2020.104281

Lake, J. A., Henderson, E., Clark, M. W., & Matheson, A. T. (1982). Mapping evolution with ribosome structure: Intralineage Constancy and Interlineage variation. *Proceedings of the National Academy of Sciences of the United States of America, 79*(19), 5948–5952. https://doi.org/10.1073/pnas.79.19.5948

Lake, J. A., Henderson, E., Oakes, M., & Clark, M. W. (1984). Eocytes: A new ribosome structure indicates a kingdom with a close relationship to eukaryotes. *Proceedings of the National Academy of Science, 81*(June), 3786–3790. https://doi.org/10.1073/pnas.81.12.3786

Lake, J. A., & Rivera, M. C. (1994). Was the nucleus the first endosymbiont? *Proceedings of the National Academy of Sciences, 91*(8), 2880–2881. https://doi.org/10.1073/pnas.91.8.2880

Landecker, H. (2017). 9. Metabolism, autonomy, and individuality. In S. Lidgard & L. K. Nyhart (Eds.), *Biological individuality: Integrating scientific, philosophical, and historical perspectives* (pp. 225–248). University of Chicago Press. https://doi.org/10.7208/9780226446592-010

Lane, N. (2011). Energetics and genetics across the prokaryote-eukaryote divide. *Biology Direct, 6*, 35. https://doi.org/10.1186/1745-6150-6-35

Lane, N. (2014). Bioenergetic constraints on the evolution of complex life. *Cold Spring Harbor Perspectives in Biology, 6*(5), a015982. https://doi.org/10.1101/cshperspect.a015982

Lane, N. (2015). *The vital question: Energy, evolution, and the origins of complex life*. Norton & Company.

Lewontin, R. C. (1970). The units of selection. *Annual Review of Ecology, Evolution, and Systematics, 1*, 1–18. https://doi.org/10.1146/annurev.es.01.110170.000245

Lindmark, D. G., & Müller, M. (1973). Hydrogenosome, a cytoplasmic organelle of the anaerobic flagellate Tritrichomonas Foetus, and its role in pyruvate metabolism. *The Journal of Biological Chemistry, 248*(22), 7724–7728.

López-García, P., & Moreira, D. (2020). The Syntrophy hypothesis for the origin of eukaryotes revisited. *Nature Microbiology, 5*(5), 655–667. https://doi.org/10.1038/s41564-020-0710-4
Margulis, L., Dolan, M. F., & Guerrero, R. (2000). The chimeric eukaryote: Origin of the nucleus from the Karyomastigont in amitochondriate Protists. *Proceedings of the National Academy of Sciences of the United States of America, 97*(13), 6954–6959.
Martin, W., & Koonin, E. V. (2006). Introns and the origin of nucleus-cytosol compartmentalization. *Nature, 440*(7080), 41–45. https://doi.org/10.1038/nature04531
Martin, W., & Müller, M. (1998). The hydrogen hypothesis for the first eukaryote. *Nature, 392*(6671), 37–41. https://doi.org/10.1038/32096
Smith, M., & Szathmáry, E. (1995). *The major transitions in evolution*. Oxford University Press.
McInerney, J. O., O'Connell, M. J., & Pisani, D. (2014). The hybrid nature of the Eukaryota and a Consilient view of life on earth. *Nature Reviews Microbiology, 12*(6), 449–455. https://doi.org/10.1038/nrmicro3271
Mereschkowski, K. (1910). Theorie Der Zwei Plasmaarten Als Grundlage Der Symbiogenesis, Einer Neuen Lehre von Der Entstehung Der Organismen. *Biologisches Centralblatt, 30*, 353–367.
Michod, R. E., & Nedelcu, A. M. (2003). On the reorganization of fitness during evolutionary transitions in individuality. *Integrative and Comparative Biology, 43*(1), 64–73. https://doi.org/10.1093/icb/43.1.64
Militello, G. (2019). Motility control of symbionts and organelles by the eukaryotic cell: The handling of the motile capacity of individual parts forges a collective biological identity. *Frontiers in Psychology, 10*, 2080. https://doi.org/10.3389/fpsyg.2019.02080
Moore, K. R., Magnabosco, C., Momper, L., Gold, D. A., Bosak, T., & Fournier, G. P. (2019). An expanded ribosomal phylogeny of cyanobacteria supports a deep placement of plastids. *Frontiers in Microbiology, 10*, 1612. https://doi.org/10.3389/fmicb.2019.01612
Moreira, D., & Lopez-Garcia, P. (1998). Symbiosis between methanogenic archaea and Delta-Proteobacteria as the origin of eukaryotes: The syntrophic hypothesis. *Journal of Molecular Evolution, 47*(5), 517–530. https://doi.org/10.1007/pl00006408
Moreno, A., & Mossio, M. (2015). *Biological autonomy: A philosophical and theoretical enquiry*. Springer.
Muñoz-Gómez, S. A., Susko, E., Williamson, K., Eme, L., Slamovits, C. H., Moreira, D., López-García, P., & Roger, A. J. (2022). Site-and-branch-heterogeneous analyses of an expanded dataset favour mitochondria as sister to known Alphaproteobacteria. *Nature Ecology & Evolution, 6*(3), 253–262. https://doi.org/10.1038/s41559-021-01638-2
Nobs, S.-J., MacLeod, F. I., Wong, H. L., & Burns, B. P. (2022). Eukarya the chimera: Eukaryotes, a secondary innovation of the two domains of life? *Trends in Microbiology, 30*(5), 421–431. https://doi.org/10.1016/j.tim.2021.11.003
Pradeu, T. (2010). What is an organism? An immunological answer. *History and Philosophy of the Life Sciences, 32*(2–3), 247–267.
Pradeu, T. (2016). Organisms or biological individuals? Combining physiological and evolutionary individuality. *Biology and Philosophy, 31*(6), 797–817. https://doi.org/10.1007/s10539-016-9551-1
Queller, D. C., & Strassmann, J. E. (2009). 'Beyond society: The evolution of Organismality'. *Philosophical transactions of the Royal Society of London. Series B, Biological Sciences, 364*(1533), 3143–3155. https://doi.org/10.1098/rstb.2009.0095
Radzvilavicius, A. L., & Blackstone, N. W. (2015). Conflict and cooperation in Eukaryogenesis: Implications for the timing of endosymbiosis and the evolution of sex. *Journal of the Royal Society, Interface, 12*(111), 20150584. https://doi.org/10.1098/rsif.2015.0584
Rivera, M. C., & Lake, J. A. (2004). The ring of life provides evidence for a genome fusion origin of eukaryotes. *Nature, 431*(7005), 152–155. https://doi.org/10.1038/nature02848
de Roos, A. D. G. (2006). The origin of the eukaryotic cell based on conservation of existing interfaces. *Artificial Life, 12*(4), 513–523. https://doi.org/10.1162/artl.2006.12.4.513
Sagan, L. (1967). On the origin of Mitosing cells. *Journal of Theoretical Biology, 14*(3), 255–274. https://doi.org/10.1016/0022-5193(67)90079-3

Santelices, B. (1999). How many kinds of individual are there? *Trends in Ecology & Evolution, 14*(4), 152–155. https://doi.org/10.1016/s0169-5347(98)01519-5

Sogin, M. L. (1991). Early evolution and the origin of eukaryotes. *Current Opinion in Genetics & Development, 1*(4), 457–463. https://doi.org/10.1016/s0959-437x(05)80192-3

Takemura, M. (2001). Poxviruses and the origin of the eukaryotic nucleus. *Journal of Molecular Evolution, 52*(5), 419–425. https://doi.org/10.1007/s002390010171

Taylor, F. J. R. (1976). Autogenous theories for the origin of eukaryotes. *Taxon, 25*(4), 377–390. https://doi.org/10.2307/1220521

Thiergart, T., Landan, G., Schenk, M., Dagan, T., & Martin, W. F. (2012). An evolutionary network of genes present in the eukaryote common ancestor polls genomes on eukaryotic and mitochondrial origin. *Genome Biology and Evolution, 4*(4), 466–485. https://doi.org/10.1093/gbe/evs018

Tovar, J., Fischer, A., & Graham Clark, C. (1999). The Mitosome, a novel organelle related to mitochondria in the Amitochondrial parasite Entamoeba Histolytica. *Molecular Microbiology, 32*(5), 1013–1021. https://doi.org/10.1046/j.1365-2958.1999.01414.x

Woese, C. R. (1987). Bacterial Evolution. *Microbiological Reviews, 51*(2), 221–271. https://doi.org/10.1128/mr.51.2.221-271.1987

Woese, C. R., & Fox, G. E. (1977). Phylogenetic structure of the prokaryotic domain: The primary kingdoms. *Proceedings of the National Academy of Sciences, 74*(11), 5088–5090. https://doi.org/10.1073/pnas.74.11.5088

Zillig, W. (1991). Comparative biochemistry of archaea and bacteria. *Current Opinion in Genetics & Development, 1*(4), 544–551. https://doi.org/10.1016/s0959-437x(05)80206-0

Zimmer, C. (2009). Origins. On the origin of eukaryotes. *Science, 325*(5941), 666–668. https://doi.org/10.1126/science.325_666

Zylstra, U. (1992). Living things as hierarchically organized structures. *Synthese, 91*(1 2), 111–133. https://doi.org/10.1007/bf00484972

Chapter 5
Living One Inside Another: Endosymbiosis, Spatial Constraints, and Functional Integration

Abstract This chapter explores the structural constraints and physiological mechanisms that enable symbiotic associations among prokaryotes to develop into physiologically integrated organisations. Specifically, it investigates the conditions that facilitate the transformation of endosymbionts into organelles, as exemplified by the evolution of mitochondria and plastids. The chapter argues that biofilms and bacterial endosymbiosis exhibit diverse spatial organisations that impose unique constraints on their constituent organisms, thereby providing distinct mechanisms of collective control. These spatial organisations confer a specific type of stability over time and open up various evolutionary pathways for converting a collective association into an integrated individual. By elucidating key physiological dimensions of functional integration, this chapter contributes to a more precise definition of "functionally integrated individuality".

Keywords Endosymbiosis · Biofilms · Spatial constraints · Biological control · Functional integration

5.1 Introduction

Chapter 3 has examined one important dimension of the functional integration of cells consisting in the interdependence between constitutive, regulatory, and signalling subsystems in prokaryotic and eukaryotic cells. This chapter, primarily based on a collaborative work done with Professor Alvaro Moreno and Doctor Leonardo Bich (Militello et al., 2021),[1] shifts our sight from individual cells to collective associations of cells. It explores the structural constraints and the physiological mechanisms that enable symbiotic associations of prokaryotes to achieve a physiologically

[1] In this chapter, I use the pronoun "I" to maintain consistency with the rest of the book, where "I" is consistently used. However, it is important to note that this chapter is the result of collaborative work, and I am deeply indebted to my co-authors. Therefore, "we" would be more appropriate in acknowledging their contributions.

G. Militello, *Functional Integration*, History, Philosophy and Theory of the Life Sciences 38, https://doi.org/10.1007/978-3-031-94744-5_5

integrated organisation. More specifically, this chapter goes to the root of eukaryogenesis by examining the enabling conditions for the transformation of an endosymbiont into an organelle as it occurred in mitochondria and plastids. As we have seen in Sect. 4.2, the appearance of mitochondria and plastids enabled the proto-eukaryotic cell not only to perform more complex functions but also to undergo key evolutionary changes that will be addressed in the following three chapters. Furthermore, the transformation of an endosymbiont into an organelle furnishes some important clues as to the philosophical issue of biological individuality, which will also be examined in Chaps. 7 and 8.

Collective associations are widespread in the biological world and give rise to very different organisations ranging from associations of bacteria and archaea to societies of multicellular organisms (e.g. social insects), yet only in certain cases these associations become an integrated individual. Thus, an intense debate about when collective associations constitute a new individuality has been taking place during the last decades. The central question underlying this debate can be summarised with Wilson's words: "at what point does a society become so well integrated that it is no longer a society?" (Wilson, 1974, p. 54). Although Wilson's question refers to *animal* societies, it is so general that it can apply to any kind of biological association. Indeed, it clearly emphasises three fundamental aspects of the transition from a biological association to an individual: first, there are some specific *conditions* (summarised by "at what point") that permit an association to become a more cohesive whole; secondly, the transformation of an association into a more cohesive whole involves the achievement of a certain degree of (functional) *integration* among the constituent organisms (indeed, they must be "well integrated"); finally, the process of integration among the parts of the association leads to something "that is no longer a society", therefore a (new) *individual* (i.e. an organismic-like associative entity).

In the contemporary debate about composite biological individuality (e.g. biofilms, holobionts, colonies of insects), the two main approaches usually adopted—i.e. *evolutionary* and *physiological*—rely on the idea that a biological individual is an integrated whole whose functions are strongly interconnected. Yet little has been said about the *conditions* that may enable an association to become a functionally integrated individual and what mechanisms are involved. The reason lies in the fact that the very concept of functional integration, often considered as a synonym for *physiological integration* (Pradeu, 2010), has not been characterised in detail. To further complicate matters, functional integration is an especially multifaceted and complex aspect of biological organisations that includes several important dimensions of a biological system such as metabolism, regulation, sensorimotor abilities, development, immunological response, reproduction, etc. Due to the unclear character of the notion of integration, not only current general definitions of biological individuality that appeal to it are somehow undermined, but also the mechanisms allowing an association to become a more integrated whole are mostly unexplored.

To develop the notion of integration in more detail with the help of biological examples, I first analyse the fundamental *physiological mechanisms* that could explain the transition from an association of bacteria towards a new full-fledged, functionally integrated individuality. Then, I examine the different *types* and *degrees* of *functional integration* enabled by different mechanisms. I aim to connect the physiological dimension of the process of individuation with the evolutionary one by analysing the role of the different forms of physiological integration in the reproduction of a collective entity.[2] Finally, I provide a *more precise characterisation* of the notion of "functional integration". It is worth pointing out that this chapter is *not* aimed at drawing up a list of properties (or criteria) that sharply distinguish a loose association of organisms from an individual, but rather at exploring the *conditions* that can potentially permit the transition from the former to the latter and, on this basis, at contributing to a better understanding of what a "functionally integrated individual" is.

Despite the huge variety of biological collective associations, I have chosen to focus on *two case-studies* from the bacterial and archaeal domains: biofilms (i.e. colonies of single- and multispecies bacteria or also archaea) and the endosymbiotic relationship between two species of bacteria.[3] These case-studies have been chosen for three reasons: first, they are *minimal forms* of composite biological systems; second, they exhibit different *physiological mechanisms* that allow understanding why the collective association achieves in each case a very different degree of functional integration among its parts; finally, the endosymbiotic relationship between two species of bacteria may provide important *clues* as to the origin of a paradigmatic example of a new functionally integrated individual, the *eukaryotic cell*,[4] by evaluating the role played by different forms of collective spatial constraints in enabling functional integration.

The connection between these case-studies lies in three main aspects: a *common spatial constraint* (i.e. the ECM and the membrane of *Tremblaya*) that surrounds a set of prokaryotic cells; the *systemic control* of parts enabled by the common boundary which affects the type and degree of physiological integration achieved by the parts; the *evolutionary potential* opened up by different kinds of common boundaries. As I shall argue in the detail, biofilms and the endosymbiosis between bacteria have diverse *spatial organisations* that constrain their constituting organisms differently, providing different *mechanisms* of *collective control*.[5] This provides them

[2] This aspect will further be analysed in Chap. 8.

[3] Other collective associations of prokaryotes include *colonies* stemming from the clones of single species bacteria/archaea (e.g. *Lactococcus lactis* or *Streptococcus thermophylus*) or intracellular parasites (e.g. *Vampirococcus* and *Bdellovibrio*). I will not analyse these cases here as they do not exhibit the features of a stable functionally integrated collective organisation. In the first case, they do not exhibit a common spatial constraint such as the EPS matrix. In the second case, intracellular parasitism is a transient not functionally integrated relationship where the host is killed.

[4] The focus on current forms of endosymbiosis as a valuable clue as to the role played by endosymbiosis in the achievement of a "strong" physiological integration in eukaryogenesis moves in a similar direction to the one explored by Reyes-Prieto et al. (2014), which focuses on non-autonomous endosymbionts with extremely reduced genomes (also called symbionelles) to shed light on the origin of eukaryotic organelles.

[5] This idea is in line with the thesis that all multicellular association need to solve the issue of spatial control, and that different ways of doing so determines different types of organisations and

with a distinct type of stability over time and opens up different evolutionary possibilities to transform a collective association into an integrated individual. This thesis has two important explanatory consequences: it sheds light on the connection between spatial constraints and physiological mechanisms for explaining the functional integration of collective associations. Furthermore, it clarifies some important physiological dimensions of the idea of "functional integration", thereby helping to better define the meaning of a "functionally integrated individuality".

The chapter is structured as follows. Section 5.2 examines some philosophical accounts that appeal to the notion of functional integration to ascribe individuality to biological associations. Section 5.3 analyses the organisation of biofilms and discusses the role of the extracellular polymeric substance (EPS) matrix in constraining and integrating the activity of the prokaryotic cells that compose them. Section 5.4 examines the organisational role played by engulfment and cross-control in the *Tremblaya-Moranella* association, the only well-studied case of endosymbiotic relationship between prokaryotes. Section 5.5 discusses the organisational differences between these two forms of association (biofilms and endosymbiosis) by focusing on the roles of the EPS matrix and engulfment. Hence, it examines the main organisational issues raised by endosymbiosis and how they may have been solved during eukaryogenesis. Finally, the last section draws some conclusions on functional integration and biological individuality.

5.2 Biological Collective Organisations: Association and Integration

The philosophical debate on biological individuality considers the concept of functional integration a decisive element to conceptualise how living systems constitute coherent wholes both in evolution and physiology. This concept is usually employed as an *explanans*, rather than the main object of investigation, as the aim is to build general accounts of individuality. In the context of the debate on evolutionary individuality, for example, Hull (1980) pointed out that, for a biological system to be an object of natural selection, it must be able not only to undergo genetic variation and transmit it to the offspring, but also to interact with the environment as a cohesive physiological whole. Moreover, integration underlies reproductive capabilities. As Sober observes, all groups (e.g. colonies of insects, groups of cells, parasitic relationships) exhibit a certain kind of functional interdependence that consists of "parts of different sorts and these parts interacts so as to sustain the organism and allow it to reproduce" (1991, p. 275). Sober suggests that all those functions involved in *self-maintenance* (e.g. metabolism) and *reproduction* somehow exhibit interdependence. Integration is also used to account for the *absence of conflict* and the presence of a *high cooperativity* among the component parts in such a way that they

different degrees of integration (Bich et al., 2019).

work as "bundles of adaptation", where all elements work toward a common evolutionary goal (Queller & Strassmann, 2009, 2016). In the same vein, Dupré and O'Malley (2009) emphasise that the functional integration of living beings, including collective associations, is characterised by the interconnection of *metabolic pathways* and *reproduction*. Metabolic processes are described generically as collaborative activities that entail a certain degree of functional interdependence. In the case of symbiotic associations, this usually takes the form of co-metabolism[6] and syntrophy.[7] Reproduction is a more complex issue in collective associations because vertical transmission and parent-offspring lineages do not always occur (see Skillings, 2016).

It is important to point out that the fundamental aim of this debate is to address the notion of *individuality* and not so much to clarify in the detail which types of *mechanisms* are required for self-maintenance, reproduction, and cooperation. Functional integration, therefore, is used as a general notion in this context.

Hence, biological individuality understood in terms of physiology has received much less attention in the literature than evolutionary accounts. In this domain, the general notion of functional integration is expected to play an even more important role. In fact, it is still employed generically. It has been emphasised, for example, that the *immune system* plays a key role in explaining the interdependence of the functional parts of organisms and collective associations, because immune interactions "are systemic (as opposed to local) and […] responsible for the acceptance or rejection of constituents in the organism" (Pradeu, 2010, p. 258; see also Howes, 1998). This view, however, does not consider the complexity of biological integration and the fact that a systemic control is performed not only by the immune system, but also by other regulatory subsystems and mechanisms that modulate and coordinate the functions of the components of the system. Moreover, the immune system depends on and is maintained by a more comprehensive physiological regime, which provides the energy for its functioning and coordinates immune activity with those of the other functional subsystems.

The philosophical approaches mentioned above rely on the notion of integration as a generic concept (mostly as an *explanans*) to develop general accounts of individuality. Instead, this chapter aims to address it as an *explanandum* by focusing on the *organisational aspects* underlying functional integration in biological associations, with the aim to identify what mechanisms enable transitions from loose collective associations to cohesive physiological wholes that are also capable of reproducing and evolving as *units of selection*. Focusing on current organisations (see Moreno & Mossio, 2015) means in the first place identifying the physiological mechanisms that enable different degrees and types of integration. This also implies identifying the possibilities and bottlenecks of different types of organisation on the evolutionary scale.

---

[6] By "co-metabolism", I mean the simultaneous degradation of two compounds: the degradation of the second compound hinges on the presence of the first compound.

[7] By "syntrophy", I refer to the phenomenon by which one species feeds on the by-products of another species.

Several works have emphasised the importance of *control* and *regulatory mechanisms* in realising integration in different types of associations, from the development of multicellular systems (Arnellos et al., 2014; Griesemer, 2016) to the physiology of symbiotic relationships (Catania et al., 2017; Bich et al., 2019).[8] Queller and Strassmann (2009, 2016), for example, have called into question the importance of spatial contiguity, the indivisibility of the parts, the development from a single lineage and the genetic uniformity among the members. They point out that it is the *control of conflict* and a *high cooperation* among the members of a society that are necessary for achieving a sufficient degree of functional integration. Interestingly, most of these accounts have focused on associations of eukaryotic organisms, which exhibit forms and degrees of integration that are different from those of the prokaryotic world. Indeed, whereas the former give rise to multicellular integrated individuals, it is doubtful whether also the latter do so.

The discussion of these ideas in the specific context of prokaryotic associations has generated a debate regarding the status of biofilms; in particular, whether (and why) they can be considered integrated individuals. Ereshefsky and Pedroso (2013, 2015) and Doolittle (2013) have recently argued that multispecies biofilms can be considered individuals, because their extracellular matrix allows for a unitary interaction with their environment and because they are capable of reproduction despite lacking a high degree of germ-soma specialisation. In response, Clarke (2016) has argued that a biofilm does not actually interact as a whole, because "most interactions take place across spatial scales that are much smaller than an entire biofilm" (p. 205). In addition, since they do not have a collective reproductive system and bacteria can enter or exit the biofilm, biofilms cannot reproduce as wholes and they cannot vertically transmit genetic variations to future generations. Therefore, they cannot undergo group selection. According to Clarke (2016), despite exhibiting a certain degree of functional cohesion resulting from metabolic co-dependence and a certain form of collective border, biofilms do not perform collective mechanisms of interaction and reproduction, and therefore they do not evolve as individuals. One may suspect that these features depend on the fact that bacterial components still keep a sufficiently high degree of autonomy and biofilms, as a system, lack more comprehensive (global) ways to control the behaviour of the bacteria.

To explore this hypothesis, I will analyse the organisational role played by the extracellular polymeric substance (EPS) matrix and other control mechanisms in biofilms. The aim is to understand how such structures and mechanisms enable a certain type and degree of functional integration in biofilms, and compare it with another type of organisation deriving from the endosymbiosis between bacteria. This latter has the potential for a different and stronger type of integration.

---

[8] As pointed out by Catania et al. (2017), *regulatory networks* play a pivotal role in defining the functional integration of symbiotic partners. These interdependent networks may be co-inherited (via vertical gene transfer) or re-established in a new generation (via horizontal gene transfer). This argument is also in line with Bich et al. (2016), who have argued that a functionally integrated organisation hinges on a complex set of regulatory mechanisms that allow it to coordinate the contributions of its functional parts and handle perturbations.

5.3 Functional Integration in Biofilms: Distributed Control and the Role of the EPS Matrix

Biofilms are ecological communities of *single-* or *multispecies* bacteria and archaea, embedded in an extracellular polymeric matrix that they produce. The development of a biofilm includes three sequential steps: the *attachment* of bacteria or archaea to a surface and the formation of a monolayer structure, which binds bacteria together and to the surface; cell division and the production of the *EPS matrix,* which gives rise to a multilayer organisation; the *disassembly* of the matrix and the dispersion of cells (Fig. 5.1).

In the first stage of biofilm life cycle, individual cells attach to a biotic or abiotic surface by means of *adhesins*[9] and give rise to a *monolayer biofilm* (Karatan & Watnick, 2009). The production of adhesins is triggered by the concentration of specific substances (e.g. oxygen or sugars) in the environment. The second stage begins when the spatial proximity of cells triggers the emission of several *extracellular signals* (e.g. mechanical, metabolic, inorganic, etc.) and the activation of *quorum-sensing* (QS) *mechanisms,* that collectively promote the synthesis and deposition of extracellular matrix components. At this stage, cells may attach to one another and to the EPS matrix, thus realising a *multilayer biofilm* where they undergo proliferation and differentiation into several cell types (Lopez et al., 2009). In the third stage, the EPS matrix disassembles and causes biofilm dispersion. This occurs in presence of a massive accumulation of toxic waste products, or when the system grows beyond the transport and distribution capabilities of EPS channels and the innermost layers of cells cannot receive enough nutrients. Biofilms employ several regulatory mechanisms that trigger dispersion in response to different stimuli (e.g. variations in concentrations of nutritional cues, oxygen and nitric oxide, presence of death bacteria) (Karatan & Watnick, 2009).

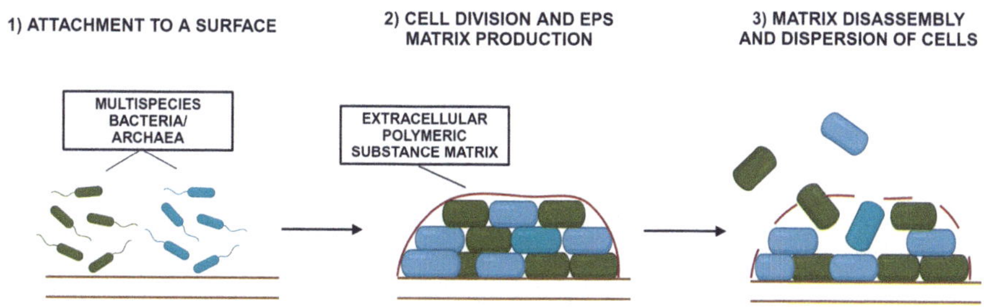

Fig. 5.1 Stages of biofilm formation and disassembly. (1) Initial attachment of multispecies bacteria and archaea to a surface. (2) Cell division and production of an extracellular polymeric substance (EPS) matrix, which stabilises the biofilm structure. (3) Disassembly of the matrix and dispersion of cells, facilitating the colonisation of new environments

[9] Adhesins are cell-surface structures of bacteria that mediate transient or permanent surface attachment.

Thanks to (at least) three types of *extra- and inter-cellular control mechanisms*—QS, EPS matrix, and bacterial conjugation—a biofilm becomes a cohesive functional unit whose parts act and are maintained together. Bacterial conjugation is somehow favoured by the close spatial proximity of cells in specific areas of the biofilm, but it works locally at *short* (cell-to-cell) *ranges*. Therefore, I will focus on the former two, which act at *medium ranges*.[10] They constitute the main factors of integration of the whole system, because they are responsible for the *overall* development and functioning of the biofilm. QS is a distributed control system that relies on the *concentrations of a* set of signalling molecules that allow bacteria to coordinate their gene expression and trigger many of the changes in the biofilm through gradients of inter cellular activation. QS is triggered when the autoinducer concentration reaches a critical threshold because of cell density (Antunes & Ferreira, 2009; Elias & Banin, 2012). It functions as a feed-forward mechanism: the bond between signalling molecules and their bacterial receptors activates the expression of several genes, including those involved in the synthesis of these same signal molecules (Saxena et al., 2019). QS plays a pivotal role in the co-aggregation of different species of bacteria in multispecies biofilm,[11] in the increase of biomass during the formation of a monolayer and a multilayer structure, and it activates many genes involved in the synthesis of matrix components (Karatan & Watnick, 2009). Finally, many bacterial species employ QS to coordinate the disassembly of the EPS matrix by promoting the inhibition of matrix components synthesis, the degradation of the matrix, and the synthesis of surfactants (Solano et al., 2014).

The EPS matrix (see Fig. 5.1.) is a dynamic structure that consists of a variety of molecules (i.e. polysaccharides, proteins, lipids, extracellular DNA (eDNA), metal ions and water), which are bound together by weak physicochemical interactions (Flemming & Wingender, 2010). The many functions of the EPS matrix—from the retention of water to enzymatic activity, from the organisation of space to protective barrier—are at the origin of the common developmental dynamics, the metabolic co-dependence, and the enhanced immunological response of biofilms.

---

[10] Short range control relies on local cell-to-cell direct interactions. Medium range control is achieved when an ensemble of cells is constrained for example by the ECM. In multicellular organisms, such as animals, it happens at the level of tissues. QS relies on signals and can be considered as a distributed medium range control mechanism because it can affect a large number of cells by generating self-organised gradients. Long range control, instead, has a systemic reach and has the potential to constrain the activity of all the parts of the system. An example of long range control mechanisms from animals is the release of hormones, distributed throughout the system through vascularisation (see Bich et al., 2019 for more details).

[11] Let us consider, for example, colonisation of the human oral cavity by the bacterial species *Veillonella atypica* and *Streptococcus gordonii*. In order to colonise dental surfaces, *V. atypica* requires the presence of *S. gordonii*, because *S. gordonii* ferments sugars and releases lactic acid, which constitutes the preferred carbon source for *V. atypica*. The co-aggregation of bacteria from the two species is made possible by the fact that *V. atypica* produces a soluble chemical signal that triggers amylase expression in *S. gordonii*, thereby increasing the degradation of complex carbohydrates and lactic-acid production (Keller & Surette, 2006).

Through mechanical forces and concentrations of eDNA, extracellular signals and enzymes, the EPS matrix places several functional constraints on the cells of the biofilm and it actively contributes to the realisation and functioning of the overall organisation of the system (see Bich et al., 2019). It makes the association of cells much more cohesive and coordinated than in the planktonic state, leading to a three-dimensional architecture (Steinberg & Kolodkin-Gal, 2015). During biofilm development, the presence of the EPS matrix mechanically inhibits the rotation of the flagella of the cells, and triggers intracellular signal cascades that increase the production and deposition of matrix molecules (Cairns et al., 2014). The "activated matrix" (Flemming et al., 2007)—characterised by the presence of digestive enzymes, signal molecules, eDNA, lytic enzymes, etc.—is involved in the exchange of genetic material, in the control of cell behaviour, in the differentiation of cells into persister cells, spores, protease cells (Cairns et al., 2014; Steinberg & Kolodkin-Gal, 2015), and in the control of the mobility of bacteria (Steinberg & Kolodkin-Gal, 2015).

The EPS matrix promotes also the spatial proximity of cells and is responsible for the presence of extracellular enzymes that give rise to an external digestive system, thus favouring integrated co-metabolism and syntrophy among symbiotic partners (Dragoš & Kovács, 2017). Moreover, fluids can flow throughout the biofilm by virtue of channels, realised by the EPS matrix, that allow the diffusion of nutrients to the cells of the innermost layers, and the distribution and removal of metabolic products (Sutherland, 2001), also enabling medium range interaction and communication within the system.

The EPS matrix allows for the formation of different biochemical environments in such a way that otherwise incompatible bacteria (e.g. aerobic and anaerobic) may co-exist in the same biofilm. EPS matrix also reduces diffusion rates of the compounds within the biofilm matrix itself, modulates gene expression patterns and decreases growth rates of the biofilm cells, making the biofilm robust with respect to external sources of perturbations and pathogens. In addition, the EPS matrix allows for the interconnection of *innate* and *induced* resistance factors that make the overall biofilm more resistant to external agents (Anderson & O'Toole, 2008), and it favours *multicellular strategies* and a *multilayer structure* that inhibit the diffusion of antimicrobial agents within the biofilm (Stewart & William Costerton, 2001). As a result, biofilms achieve a certain form of collective *immunological* capability and gain a fitness advantage over their planktonic state (Burmølle et al., 2014).

In nature, multispecies biofilms tend to be more common than single-species biofilms. However, these latter are present in a variety of infections and on the surface of medical implants (Bermejo et al., 2019). Single- and multispecies biofilms are essentially similar both in the stages of the extracellular matrix deposition and degradation (attachment, maturation, and dispersion) and in the mechanisms involved in bacterial communication (quorum sensing) and bacterial conjugation. Nevertheless, they exhibit some differences in the interactions among bacterial partners. Multispecies biofilms exhibit much more variety of exchanges of nutrients and electrons than single-species biofilms, and thus they "gain energy from a series of reactions that a single species might lack" (Lohse et al., 2018, p. 27). Furthermore, it has been observed that the inclusion of other bacterial species in a single-species community may provide their members with numerous *physiological advantages* (e.g. passive resistance, metabolic cooperation, more efficient DNA sharing) (Wolcott et al., 2013). It is worth noting that, compared to single-species biofilms, multispecies biofilms can develop both *cooperative* relationships leading to increased biomass of the bacterial members and *competitive* relationships producing a decreased biomass of all members (Liu et al., 2016). It therefore seems that the life cycle of multispecies biofilms is subject to sharper fluctuations and, therefore, variability than that of single-species biofilms, thus potentially providing multispecies biofilms with increased capacities to invade surfaces, proliferate, and develop drug resistance. In the light of all these characteristics, it is sensible to suggest that both single- and multispecies biofilms exhibit the *same kind* of physiological integration enabled by the extracellular matrix; nonetheless, internal differentiation and functional diversity seems to be higher in some types of multispecies biofilms because of the higher variety of metabolic exchanges between bacterial partners.

In the light of the above, what *type* and degree of integration—and therefore, of individuality—does this form of association achieve? Integration in biofilms is achieved by *means of collective control* exerted by QS mechanisms and EPS matrix at longer ranges than those that characterise basic cell-to-cell interactions alone. This is also a *coarse-grained type control*, based on differences in concentrations (of signals, control macromolecules, etc.), which is exerted gradually in space, at short and medium ranges, through gradients of signalling interactions (QS) and of distributions of EPS molecules. Although this form of control is *not specific*—i.e. does not rely on single interactions for a certain effect—it can give rise to a high variety of behaviours within the collective system.

From an organisational perspective, the EPS matrix structures are higher-level control subsystems exerted over the individual cells. They contribute at medium ranges to the structural and functional cohesiveness and cooperation within biofilms. This is the reason why some authors have regarded biofilms as interactively and evolutionarily cohesive biological integrated individuals (Ereshefsky & Pedroso 2013, 2015; Doolittle, 2013), and even as full-fledged multicellular *organisms* (Shapiro et al., 1997). However, due care should be exercised with regard to the *type* and *degree* of functional integration and individuality of biofilms, inasmuch as the capability of the EPS matrix to give rise to a fully integrated system is *limited* by several factors. Firstly, the EPS matrix does control the activity and fate of their cells only at short and medium ranges, but not at long ranges (relative to the scale of the system), due to the lack of full-fledged vascularisation, among other things (Bich et al., 2019). It is worth noting that there is no specific constraint that has the capability of exerting a long-range type of control with a systemic reach. Indeed, global effects are achieved through self-organisation processes that are the result of coarse-grained distributed control such as chemical gradients. Secondly, since the EPS matrix enables some degree of spatial segregation and functional differentiation only by means of gradients of concentrations, the internal modularity of the system is limited. Thirdly, the EPS matrix lacks components that make possible modularity and the construction of a global interface with the environment.[12]

In sum, while providing cohesiveness, the EPS does not establish *clear-cut global boundaries* or *interfaces*, nor *long range control mechanisms*. Therefore, if compared with specialised membrane mechanisms found in unicellular systems or interfaces such as the epithelium in eukaryotic multicellular systems, the EPS exerts a weaker and less specific control upon the permeability and selectivity of the system as a whole. Furthermore, the modulation of fluid transport by EPS channels is limited. This in turn limits the overall capability of the collective system to grow and to control cells at longer ranges.

The organisation of biofilms allows their cells to keep their autonomy and revert cell differentiation. Although some specialised cells may play the function of spores (e.g. in *Bacillus subtilis* biofilm) (Claessen et al., 2014) and some cheats (e.g. in *Pseudomonas aeruginosa* biofilm) are considered as primitive forms of a germ cells (Rainey & Kerr, 2010; Hammerschmidt et al., 2014), this type of differentiation—which is the result of self-organisation starting from local interaction rather than specific control mechanisms—does not satisfy the requirements for units of selection. Moreover, each germ cell has its own history of mutation as a soma cell before randomly differentiating into a reproductive spore. In addition, most biofilms are characterised by the entrance and dispersion of cells, so that it can be claimed that biofilms do not exhibit a type of reproduction coordinated at the level of the whole system. The lack of a unified reproduction is even more evident in multispecies biofilms, where different genetic pools are represented without a reproductive bottleneck.

The fact that biofilms lack standard reproductive criteria for individuality (e.g. high levels of germ-soma specialisation, unified reproductive lineages, reproductive bottlenecks) poses some challenging questions about whether they can be regarded as units of selection, and thus evolutionary individuals. This issue forms the core of the debate between Clarke's (2016) and Ereshefsky and Pedroso's accounts (2013, 2015). According to Ereshefsky and Pedroso (2013, 2015), the criterion of evolutionary individuality based on the transference of genes from parents to offspring (vertical transmission) within the same lineage is too narrow. Thus, they propose a more open-ended approach according to which the members of a prokaryotic association (e.g. the prokaryotes of multispecies biofilms and consortia) share genes that provide them with mechanisms for trait transmission and reproduction. Thus, they "achieve evolutionary individuality but do not transmit their traits through single-species lineages. [...] Trait transmission in such consortia is accomplished through both lateral and vertical gene transfer, and the reproduction (or production) of such consortia is typically accomplished by aggregation" (Ereshefsky & Pedroso, 2015, p. 10131). By contrast, Clarke (2016), who defends a view of heritage and evolutionary individuality based on parent-offspring lineage, has argued that the EPS

[12] For example, the EPS matrix lacks collagen IV, which promotes the realisation of interfaces and organ formation in eukaryotic multicellular systems, due to the role it plays in the basement membranes (Fidler et al., 2017).

matrix cannot give rise to a common lineage, and therefore it is not possible to regard biofilms as units of selection, since heritable variation occurs at the level of the single bacterial components rather than at the level of the biofilm as a whole multispecies.

5.4 Endosymbiosis in Prokaryotes: Engulfment and Its Implications

In this section, I analyse a very different form of prokaryotic association, based on asymmetric engulfment of a species of bacteria within another (i.e. endosymbiosis). I am interested in this form of association because it is presumably of the type that led to a paradigmatic case of highly functionally integrated system: the eukaryotic cell and its organelles of endosymbiotic origin. It is however worth noting that the endosymbiotic events that led to the origin of mitochondria and plastids in the eukaryotic cell were extremely rare in the prokaryotic world (Lane, 2005; Booth & Ford Doolittle, 2015), probably because of the difficulty of overcoming conflicts between two prokaryotes fostering an endosymbiotic relationship.

Indeed, only *one* case of evolutionary stable[13] endosymbiotic relationship between two prokaryotes has been discovered so far[14]: a γ-proteobacterium (*Candidatus Moranella endobia*) that lives inside a ß-proteobacterium (*Candidatus Tremblaya princeps*). This association is very peculiar because it is not capable of an independent form of life; indeed, it exists only enclosed in specialised cells (the bacteriocytes) of a specific organ (the bacteriome) of the mealybug insects[15] (Fig. 5.2). Phylogenetically, the symbionts entered the mealybug at different times, the *Tremblaya* first and the *Moranella* later, so that their endosymbiotic relationship originated within the insect. In this sense, this endosymbiotic association shares some organisational features with the prokaryotic endosymbionts of sap-feeding insects, although in these latter cases the prokaryotic organisms live in eukaryotic

[13] By "*evolutionary* stable relationship", I mean a relationship that persist across several generations and that undergoes natural selection as a whole.

[14] Intracellular bacteria have been identified in some blue-green algae of the species *Pleurocapsa minor* in the seventies (Wujek, 1979), but the physiology of this association has not been investigated. Other cases of intracellular bacteria invading the periplasm (e.g. *Bdellovibrio*) or the cytoplasm (e.g. *Daptobacter*) of other bacteria have been found (Corsaro & Venditti, 2006). However, these cases represent *transient* symbiotic relationships (i.e. parasites) that do not give rise to *an evolutionary stable* relationship.

[15] For the sake of the argument, I just focus on the endosymbiotic relationship between the two bacteria (*Tremblaya* and *Moranella*), leaving aside the functional contribution of the insect. Indeed, this chapter studies the functional integration of associations of *prokaryotes* and not the functional interdependence between prokaryotes and (multicellular) eukaryotes. Therefore, for clarity, hereinafter I will use the term "host" to refer to *Tremblaya*, whereas the term "endosymbiont" refers to *Moranella*. I will use "mealybug cells" for those eukaryotic cells that contain the *Tremblaya-Moranella* association.

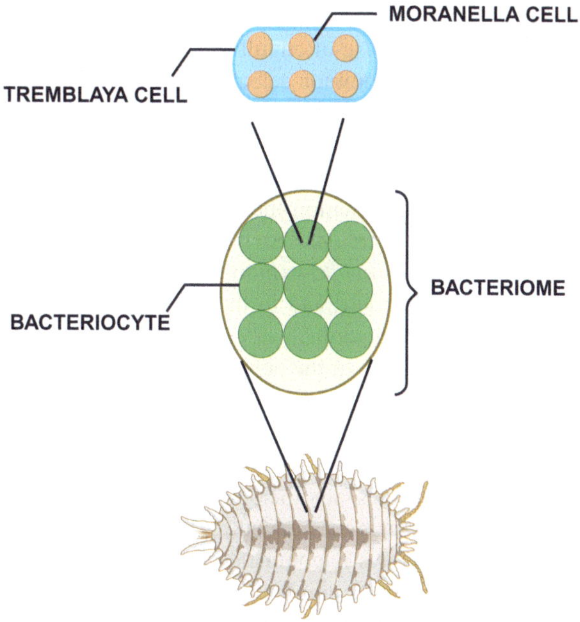

Fig. 5.2 The endosymbiotic relationship between *Tremblaya princeps* and *Moranella endobia*. *Moranella* resides within *Tremblaya*, forming a nested symbiosis. Both symbionts are located within bacteriocytes, specialised cells of the bacteriome in mealybug insects (*Planococcus citri*)

cells (von Dohlen et al., 2001; McCutcheon & von Dohlen, 2011). Like many endosymbionts of sap-feeding insects (e.g. *Hodgkinia*, *Carsonella*, *Sulcia*, etc.), both *Tremblaya* and *Moranella* have extremely reduced genomes that affect their metabolic (i.e. anabolic and catabolic pathways), genomic (i.e. DNA replication, transcription and translation), and regulatory functions (i.e. metabolic regulation and gene regulation). *Moranella*'s genome, which is four times larger than *Tremblaya's*,[16] codes for RNA molecules and proteins that cannot be expressed by *Tremblaya* genome; but *Moranella*'s genome is far from being self-sufficient and the functioning of this organism depends, in turn, on some of the few gene products from *Tremblaya*.

Despite these specificities and the impossibility to consider the consortium as a living fossil of an earlier step in the eukaryogenesis, the *Moranella-Tremblaya* association deserves a careful analysis for my purposes, precisely because it shows the early-stage implications of engulfment among two types of prokaryotic cells. Thus, it can shed light on the possibilities opened by this relationship and on the organisational problems it needs to overcome in order to maintain viability. Let us examine how engulfment affects this symbiotic relationship.

---

[16] *Tremblaya*'s genome is 138,927 bp in length, whereas *Moranella*'s is 538, 924 bp (McCutcheon & von Dohlen, 2011). The difference in genome size between *Tremblaya* and *Moranella* is consistent with the hypothesis that *Moranella* penetrated *Tremblaya* as a secondary endosymbiotic event (López-Madrigal et al., 2013a).

Engulfment leads to a situation distinct from the one created by the EPS matrix in biofilms. In the case of engulfment, the host, *Tremblaya*, and the endosymbiont *Moranella* share a common selective boundary, defined by *Tremblaya*'s membrane, which coordinates their different metabolic organisations. This is a global constraint which enables a systemic long range control on all parts, and determines the type and degree of physiological integration between them. Moreover, since one of the organisms is located within the cytoplasm of the other, global viability requires a different type of functional coordination. Like many endosymbiotic relationships, *Tremblaya* and *Moranella* exhibit *metabolic complementation*, since the symbiotic partners partially contribute to the same metabolic pathways. A good example is provided by the metabolism of carbohydrates. *Tremblaya* has the genes encoding for only two enzymes of the pentose phosphate pathways (transaldolase B and transketolase). The rest of the enzymes for the pentose phosphate pathways, glycolysis, the phosphotransferase system, and the pyruvate dehydrogenase complex are expressed by *Moranella's* genome (López-Madrigal et al., 2013a). Amino acid biosynthesis constitutes another clear example of metabolic complementation: *Tremblaya* contains genes encoding for ten essential amino acids, but none of the amino acid pathways is complete in either *Tremblaya* or *Moranella,* so that these pathways need to be complemented by a patchwork of metabolites and enzymes from both partners (McCutcheon & von Dohlen, 2011). Other metabolic pathways are incomplete (e.g. the tricarboxylic acid cycle) or absent (e.g. the nucleotide synthesis de novo or the synthesis of vitamins and cofactors) in the consortium.

The functional interdependence exhibited by the *Tremblaya-Moranella* association, however, is much deeper than metabolic complementarity and complementation—i.e. the exchange of metabolites or intermediate substrates, respectively—which are widespread in nature and characterise biofilms as well. Importantly, *Tremblaya* and *Moranella* jointly realise the iron-sulphur [Fe-S] cluster,[17] which is usually not fully preserved in endosymbionts with reduced genomes (López-Madrigal et al., 2013a). The synthesis and assembly of this cluster requires a complex molecular machinery, and both members of the consortium are involved in the synthesis and maintenance of it, thus exhibiting a high degree of coordination. Another important aspect is that *Tremblaya* is totally dependent on *Moranella* for ATP synthesis, a feature that probably makes this consortium, according to Lopez-Madrigal et al. (2013a), the only known case in which all energy sources appear to be provided by only one of the partners. This is somehow analogous to what happens in the eukaryotic cell, where the mitochondria perform this function. Additionally, the cell-envelope structure is simplified in both bacteria, because both *Tremblaya* and *Moranella* have lost most of the genes for the synthesis of murein and lipopolysaccharides.

[17] The [Fe-S]-cluster is a prosthetic group mainly involved in oxidation-reduction reactions. It plays several important functions related to energy metabolism and regulation. Notably, it plays a role in bacterial (and mitochondrial) respiratory complexes, in enzyme catalysis and in the sensing environmental or intracellular conditions to regulate gene expression (Lill, 2009).

The high degree of functional coordination between the endosymbiont and the host can be observed in the way how their *genetic functions* are realised.[18] Transcription requires the contribution of both organisms. *Tremblaya* encodes all the essential subunits of RNA polymerase and a single sigma factor, but it lacks the genes responsible for the basic transcription machinery, and for RNA processing and degradation. By contrast, *Moranella* has a minimal but complete transcription machinery and a number of genes encoding proteins that assist transcription. Furthermore, several transcriptional regulators, the functions of which are not yet fully known, and which are usually absent in endosymbionts with reduced genomes, have been retained by the genome of *Moranella* and they may play a role in the control of the transcription in this organism. Regarding translation, the consortium performs a very complex *functional complementation*[19] and, according to López-Madrigal et al. (2013a), it may constitute the only known case for this specific function. While *Moranella* encodes more than 80% of the tRNA genes for the consortium, *Tremblaya* has retained tRNA genes for the most frequently used codons for alanine and, importantly, those for lysine, which are missing from *Moranella's* genome. Both *Tremblaya* and *Moranella* code for a high number of ribosomal proteins, giving rise to a ribosomal redundancy that could play a (not yet known) functional role for both symbiotic partners. However, only *Moranella* encodes ribosome maturation proteins and translational factors. In sum, the consortium shows a high degree of genetic complementarity: *Tremblaya* has lost most of the genes not only for metabolic but also for genomic functions, whereas *Moranella* has retained different genes for metabolism and genomic functions that *complement* those of *Tremblaya*, giving rise to a highly co-dependent relationship in which each partner partly contributes to the control mechanisms of the consortium.

In the light of the above, a fundamental question arises: how can this high degree of functional complementation happen? To answer to this question, I shall examine how *Tremblaya* and *Moranella* share functional constituents such as proteins through *Moranella*'s membrane. To perform its essential cellular functions, *Tremblaya* needs to import from *Moranella*'s cytoplasm not only metabolites, amino acids or carbon sources—as it is typical in common cases of endosymbiosis—but also functional control components such as proteins, tRNAs, ATP and molecular complexes (McCutcheon & von Dohlen, 2011; López-Madrigal et al., 2011, 2013a).[20] The consortium, therefore, requires a special transport system for the exchange of big molecules between the two partners.[21] *Moranella*'s genome

[18] See López-Madrigal et al. (2013a) for the details.

[19] By "functional complementation", I mean the exchange of components that perform, or contribute to, specific functions (such as proteins, tRNA, parts of ribosomal machinery etc.) between the members of the association. It is different from "metabolic complementation" which, instead, consists in the exchange of intermediate metabolic substrates.

[20] The same happens, in lesser degree, in the opposite direction from *Tremblaya* to *Moranella*.

[21] Exchanging small molecules such as amino acids and metabolites is deeply different from exchanging proteins, tRNA or other big molecules, and the two cases depend on distinct mechanisms of transport. Unlike the *Tremblaya-Moranella* relation, the endosymbiotic relationship

encodes a limited set of active transporters (e.g. the phosphotransferase system for the transport of hexoses) and two channels (MscL and YbaL) associated with osmotic stress, which play an important role in the excretion of low molecular weight molecules (e.g. ions and metabolites) and small cytoplasmic proteins. The Sec translocon[22] machinery of *Moranella* exhibits a very reduced protein permeability and, therefore, it does not seem to be responsible for the provision of proteins and RNAs to *Tremblaya*. Hence, it has been hypothesised that the protein translocation from *Moranella* to *Tremblaya* may be due to a very primitive mechanism, yet effective in this case (López-Madrigal et al., 2013b). It would consist in a transient perforation of the *Moranella* plasma membrane and of the osmotic channel MscL, controlled by osmotic stress.[23] This transient perforation is made possible by two factors: (1) the peculiar composition of *Moranella*'s membrane, more subject to perforation; and (2) the unequal distribution of metabolic products in the two partners. In this way, the cell wall of *Moranella* is transiently damaged and proteins would be able to reach *Tremblaya* cytoplasm (López-Madrigal et al., 2013b), thus allowing the two partners to exchange control components such as proteins and operate in an integrated way.

From the above, we can make some important remarks. The establishment of an *endosymbiotic relationship* between two bacteria is a condition that has rarely been observed in the biological world, because engulfment creates an intimate relationship within the cytoplasm of one of the partners that is difficult to maintain. Indeed, the absence of a compartmentalised nucleus makes the genome of the prokaryotic host, not protected by the nucleus membrane, more susceptible to a bombardment from pieces of DNA of endosymbiotic origin, in such a way that the host would be genetically unstable and, therefore, not adaptive (Lane, 2015). This is one of the reasons why bacterial engulfment usually takes the form of a *transient* and *ephemeral* parasitic relationship, which ends when the prey (the "host") is killed (e.g. *Vampirococcus* and *Bdellovibrio*) or the predator leaves the prey (e.g. *Daptobacter*).

The endosymbiotic relationship between *Tremblaya* and *Moranella* shows the exceptional conditions required nowadays for a viable mutualistic relation. The massive loss of genes,[24] as demonstrated by *Tremblaya*'s genome, implies that many

between a eukaryotic host and prokaryotic endosymbionts usually relies only on the exchange of amino acids between the two partners. The import of amino acids into the endosymbionts and the export of other amino acids to the host is usually mediated by transporters provided by the host (Duncan et al., 2014).

[22] A translocon is a complex of secretory (sec) proteins involved in the translocation of polypeptides across the membranes.

[23] Osmotic stress (or shock) is a sudden change in the solute concentration around a cell causing a change in the movement of water across the cell membrane.

[24] The loss of genes is an important feature of many commensal and mutualistic (symbiotic) relationships and has been hypothesised to increase the fitness of the overall associations. Morris et al. (2012) have coined the expression of "Black Queen hypothesis" to posit that "certain genes, or more broadly, biological functions, are analogous to the queen of spades. Such functions are costly and therefore undesirable, leading to a selective advantage for organisms that stop performing them. At the same time, the function must provide an indispensable public good, necessitating its

metabolic pathways are absent or incomplete (e.g. those for amino acid biosynthesis). More importantly, some genetic functions (e.g. DNA replication, recombination, repair, and transcription), and regulatory processes (e.g. transcriptional regulators and translational factors) are severally undermined. As we have seen, most of these functions are either complemented or supplied by *Moranella* and thus require the joint action of both partners. This suggests a *highly functional interconnected organisation* in which each bacterium requires the other one. For these reasons, Lopez-Madrigal and colleagues make the strong claim that "'*Ca. Tremblaya princeps*' cannot be considered an independent organism, but that the consortium with its gammaproteobacterial symbiotic associate represents a new composite living being" (López-Madrigal et al., 2011, p. 5587).

This case study is not aimed at putting the consortium in the same category as, for example, a eukaryotic cell, characterised by several specialised mechanisms that control the interaction between the cell and its organelles of symbiotic origin. I rather suggest that the *engulfment* between prokaryotes poses a series of challenges that need to be overcome in order to maintain a viable association. Unlike in bacterial biofilms, the establishment of a common boundary (the membrane of *Tremblaya*) makes possible the control of the global boundary conditions of the *Tremblaya-Moranella* system. While in biofilms the EPS matrix allows medium range control upon ensembles of cells, the endosymbiosis realised by the *Tremblaya-Moranella* association depends on a global spatial constraint, the membrane of *Tremblaya*, which has a (long-range) systemic control upon all the components (the molecules and the endosymbionts in the cytoplasm of *Tremblaya*). Another important type of spatial constraint is the membrane of the *Moranella* endosymbionts. The presence of an internal spatial constraint, the membrane of *Moranella* symbionts, allows for a further compartmentalisation and modularity with the potential for the evolution of specific controllers capable to modulate the permeability of the internal compartment, like what happened in the case of organelles during eukaryogenesis. While exerting a systemic constraint at the global level, the endosymbiotic association exhibits more specific forms of molecular control. This is achieved through not only self-organisation, but also functional components from either *Tremblaya* or *Moranella*, or functional components which are assembled from parts synthesised by both symbionts.

Moreover, the engulfment favours the genetic reduction and functional reorganisation of the two symbionts, thus leading to symbiotic partners that are necessarily less autonomous than those of a biofilm. A fundamental aspect of the integration between *Tremblaya* and *Moranella* is that they exchange not only metabolites and amino acids, but also their main functional components. In the stable context of a

---

retention by at least a subset of the individuals in the community" (Morris et al., 2012, p. 3). In most cases, symbiotic partners share metabolic products, giving rise to forms of "syntrophic" integration that contribute to establish ecological networks (Skillings, 2019). In other cases, the members of the association share functional components under collective constraints such as EPS matrix or a common boundary, thus giving rise to forms of cross-control that allow for forms of physiological integration (Bich et al., 2019).

nested endosymbiosis within the mealybug, this consortium has provided primitive, yet effective responses to the aforementioned challenges (e.g. the presumed passage of proteins, tRNA, etc.) through mechanisms related to osmotic stress.

5.5 EPS Matrix and Engulfment: A Comparative Discussion

The previous two sections have highlighted the distinctive organisational features of biofilms and the endosymbiosis between bacteria: their different types of collective borders that spatially constrain at different ranges the members of these associations and the forms of controls they enable, from coarse-grained distributed control in biofilms to more specific fine-grained ones shared by both partners in the endosymbiotic association. In this section, I compare their systemic implications. Based on this comparative discussion, I investigate the conceptual links between the engulfment of the *Tremblaya-Moranella* association and that of mitochondria and plastids in the eukaryotic cell.

In the case of biofilms, the role of the collective spatial constraint is played by the EPS matrix: a dynamic extracellular structure that provides global cohesion, controls the activity of whole groups of bacteria and archaea at short and medium ranges, and maintains the cells adjacent to one another while differentiating gradients of space characterised by different boundary conditions. This allows for functional differentiation and co-metabolism, syntrophy, common development and an enhanced immune response of the overall biofilm. In the case of the endosymbiosis between two bacteria, instead, engulfment provides the two organisms with a common global membrane (i.e. the membrane of *Tremblaya*). This is a global spatial constraint that favours an asymmetric relation between the host and the endosymbiont with a much more intimate and demanding collaboration between the members of the association, which forces the establishment of systemic long-range control (relative to the size of the whole system) and make it possible the realisation of fine-grained specific control mechanisms to coordinate the activities of the members.

The EPS matrix of biofilms and the membranes of the endosymbiotic associations impose on the symbiotic partners *different types of spatial constraints* that affect their collective physiological functions. Although both EPS matrix and engulfment allow for metabolic coordination, they lead to *different* forms of collective *metabolic* organisations. In the case of biofilms, the spatial proximity of bacteria within the EPS matrix favours co-metabolism and syntrophy, and the release of enzymes in the EPS gives rise to an external digestive system. Nevertheless, bacteria keep their autonomy and can in principle develop new metabolic relationships or leave the biofilm altogether. In the case of endosymbiosis, by contrast, *engulfment* implies a very specific set of selective pressures that pave the way for a symbiotic "rabbit hole" in which incomplete metabolic pathways of the host are complemented by those of the endosymbiont and vice versa (Bennett & Moran, 2015). Accordingly, in this inescapable association, metabolic interdependences are not facultative nor easily realisable by interchangeable partners, and they are also much more stable across time (i.e. over generations) if compared to the case of biofilms.

Yet the difference between bacterial endosymbiosis and biofilm is not limited to the stability of metabolic interdependencies, but includes two other aspects that have deeper organisational implications. In the first place, the bacterial EPS represents a collective and fuzzy spatial constraint, with little and *unspecific control* upon the passage of molecules or organisms through it. In the case of engulfment, instead, the membrane of the host provides the system with a common selective border characterised by *global and more precise mechanisms* of control of permeability and transport (i.e. the capability to modulate the internal pH, concentrations of metabolites, osmotic pressure, spatial and temporal distribution of specific chemicals). With regard to the internal organisation of space, the EPS allows for a *coarse-grained* spatial differentiation, while the membrane of the endosymbionts provides the endosymbiotic association with much more modularity due to the presence of internal compartments, opening up the possibility for a *fine-grained* control of permeability of the endosymbiont membrane and targeted transport of proteins.

In the second place, there are fundamental differences between the two types of associations in the way how they control *intercellular relations*. In biofilms, QS and the EPS matrix exert a *distributed control* upon the cells at short and medium ranges through gradients of signalling interactions (QS) and distributions of EPS molecules. In engulfed symbiosis, instead, functional coordination requires *specific control* at all ranges and avoidance of conflict. In the case of the *Moranella-Tremblaya* association, for instance, proteins, tRNA and other control molecules of *Moranella* need to pass across its membrane to enter the cytoplasm of *Tremblaya* where they can directly control different biosynthetic processes or regulate basic functions. To a lesser degree, the same happens to some control molecules from *Tremblaya*, so that the whole consortium maintains viability through a very basic form of cross-control[25] and interlocked regulation.[26] Thus, engulfment cannot succeed unless a tight control of the most fundamental functions of both the host and the endosymbiont is established. This requires a deep functional re-organisation with an irreversible loss of autonomy of the former partners.

For these reasons, I suggest that the engulfment between two prokaryotes constitutes the fundamental requirement for the appearance of a strong, and non-facultative, *functional integration* between different symbiotic partners. This type of relationship is very demanding in organisational terms, insofar as such a specific control requires: (1) the presence of the right components in the right place at a given time; and (2) the implementation of mechanisms for transporting proteins and

---

[25] By "cross-control", I mean one partner producing the components that control processes in the other.

[26] By "interlocked regulation", I mean the activity of regulatory mechanisms which rely on the components produced by both partners.

other complex control macromolecules across the membrane of the endosymbiont, not only basic building blocks such as metabolites and amino acids.[27]

These different types of physiological integration pose some difficult questions about the relationship between functional integration, system-level coordinated reproduction, and heredity. It is not my purpose to find a solution to this complex issue; however, I suggest that in collective organisations a certain level of physiological integration is required to gain the capability of reproducing as a unit,[28] because parts need to be functionally differentiated (e.g. between germ and soma) and their activities coordinated for this purpose. For example, under starvation conditions, some single-species biofilms (e.g. *B. subtilis* or *M. xanthus*) can produce collective forms of reproduction (e.g. the spores of *B. subtilis* or the fruiting bodies of *M. xanthus*) by means of (local) *contact mediated signals*, that are enabled by the spatial proximity of cells, and the resulting formation of gradients through self-organisation (Julien et al., 2000; Muñoz-Dorado et al., 2016). Thus, the existence of *extracellular matrix*, which keeps bacteria close to one another, may play an organisational role not only in establishing a certain kind of physiological integration, but also a *diffused* and *transient*, *context-dependent* (i.e. dependent on starvation conditions), collective reproductive system. This is realised through coarse-grained control, based on local interactions and gradients, rather than coordinated by means of global specific control mechanisms. Similarly, some "multicellular prokaryotes" (Claessen et al., 2014; Lyons & Kolter, 2015) such as filamentous bacteria (e.g. *N. punctiforme*), actinomycetes (e.g. *A. Israeli*) or beggiatoa (e.g. *B. leptomitoformis*) exhibit distinct kinds of collective reproduction that seem to be enabled by their spatial contiguity (e.g. through proteinaceous complexes) and a minimal degree of functional integration (e.g. intercellular signals, metabolic co-dependence) (see Claessen et al., 2014). In the case of endosymbiosis, *engulfed symbionts* are so tightly integrated that they cannot survive autonomously, and the endosymbionts can only be transmitted vertically due to the role of the host's membrane as global constraint. Consequently, the genes of both the host and the endosymbionts jointly change, and these variations can selectively be transmitted to the new generations, *making the whole system a unit of selection*. Moreover, engulfment allows for the implementation of further fine-tuned regulatory mechanisms during the evolution of the symbiosis to synchronise the processes of growth and division more precisely.

A basic form of coarse-grained physiological integration is therefore a necessary, but not sufficient, condition for a collective reproduction and a vertical transmission

---

[27] The *Moranella-Tremblaya* consortium arranges transport through a very basic mechanism based on osmosis in presence of a weakened membrane. Despite being unspecific and inefficient, it can guarantee the viability of the consortium in the very stable environment of mealybug cells. In fact, this mechanism lacks the presence of complex channels and protein targeting mechanisms that would allow much more specific control upon the localisation of functional components. A more stable and robust solution to this problem would instead require a much deeper re-organisation of the systems involved. This is what it is supposed to have happened during the process of eukaryogenesis.

[28] The relation between functional integration and system-level reproduction will further be studied in Chap. 8.

of genes, as shown by multispecies biofilms. They have an EPS matrix that allows for distributed mechanisms of regulatory control and signalling for synchronising the members of the association. However, neither a global (long-range) control upon the reproduction of the components, nor unified mechanisms for the differential variation of the gene pool of a biofilm have been reported in the current literature so far (see, for example, Lopez et al., 2009; Liu et al., 2016). As a result, the cells of single- and multispecies biofilms can evolve *independently from one another* rather than undergoing *co-selection*, as *evolutionary* individuals instead do. By contrast, the type of integration enabled by a common boundary and control constraint in engulfment paves the way not only for a higher degree of integration and functional differentiation, but also for a collective reproduction, even though the species involved are different.

The engulfed association among prokaryotes imposes a specific set of constraints on the symbiotic partners, which determine a dramatic reduction in the endosymbiont's genome, and lead to an irreversible functional dependence between the symbiotic partners. Despite its limits, the relevance of the relationship between *Tremblaya* and *Moranella* lies in that it involves not only a sophisticated complementation of metabolic and genomic functions between the host and the endosymbiont, but also control upon protein localisation and the reproduction and development of the endosymbionts. These dynamics, which are hardly sketched in the consortium *Tremblaya-Moranella*, achieve the highest expression in *the eukaryotic organelles* of *endosymbiotic origin*, such as mitochondria and plastids, to give rise to full-fledged functionally integrated systems: the eukaryotic cells. Since most of the proteins controlling and regulating the internal processes of proto-mitochondria and proto-plastids were progressively encoded in the nucleus and synthesised in the cytoplasm of the host cell, the appearance of a *protein import and targeting machinery* played a major role in the conversion of endosymbionts into organelles. It allowed the host to directly control and regulate the functions of the future organelle and the overall consortium (Cavalier-Smith, 2007; Martin, 2010).

In sum, a viable engulfment establishes several constraints on the organisations of both the host and the endosymbionts. These constraints are much tighter and demanding than those placed by the EPS matrix on the bacteria of a biofilm. It is not a coincidence, therefore, that only very few cases of prokaryotic endosymbiosis have been discovered so far. The evolutionary stable internalisation of a prokaryotic organism within another one raises a series of issues whose solution leads to a deeper reproductive, developmental, and metabolic integration, based on a more precise form of central control. Ultimately, they may give rise to the appearance of a strong integrated identity. The EPS matrix, instead, leads to more ephemeral, although highly successful, organisations, in which symbiotic partners retain a basic autonomy and are kept together by distributed forms of control, without a fine-tuned control of the overall development and reproduction.

5.6 Concluding Remarks

Our case studies show that in the prokaryotic world, a process of association of individuals may lead to different forms of collective units, with *different types and degrees of integration*. In this context, functional integration can be broadly understood as a phenomenon that originates when a set of different and initially autonomous organisations—each one with its own functional parts—begins to functionally cooperate and share their local functions. This leads to the establishment of a wider collective organisation where some functional constraints of the constituent organisations are interlocked and control one another's processes in such a way that the whole system achieves viability. In this context I have identified a series of crucial elements that enable different types and degrees of functional integration, specifically: (1) Different types and ranges of collective spatial constraints exerted by the EPS in biofilms and the global membrane in the endosymbiotic association. (2) Different forms of control exerted by both spatial and intercellular control mechanisms such as the distributed coarse-grained control of biofilms and the specific fine-tuned control of endosymbiosis. (3) The different degrees of cross-control and interlocked regulation that are required to modulate and coordinate intercellular interactions in the different associations. In *Tremblaya-Moranella* complex, interlocked regulation is extremely strong, as most of the biological mechanisms are assembled from functional components produced by both partners.

However, the necessity of the associated parts to achieve a global viability implies that there could not be an indefinite number of integration possibilities. Instead, given a specific set of entities, only a discrete number of collective organisations are stable on the physiological and evolutionary scales. In prokaryotic collective organisations, we can find either biofilm-type forms of association or endosymbiotic associations of prokaryotic cells; but the latter seem, for all the reasons that I have analysed, quite rare and fragile until a completely new organisation—the eukaryotic cell—is realised.

As we have seen, the appearance of a strong form of functional integration in the prokaryotic domain is a process characterised by an initial step where different individualities (autonomous entities) enter in a process of irreversible association. Most symbiotic associations of prokaryotes, like biofilms, cannot be considered as full-fledged functionally integrated individuals, but rather as communities of (sometimes highly) coordinated organisms, kept together by means of distributed control mechanisms. Among these associations, the key for the achievement of a *strong functional integration* is the creation of an *asymmetric compartmentalisation*: a spatial border, a global constraint that functionally acts as a selective frontier between the associated system and the external environment. At the same time, engulfment is a much more difficult way of achieving a stable association, and it was at the origin of a long evolutionary travel full of conflicts.

Engulfment triggers a cascade of events that opens up several possibilities for structural and functional reorganisation and biological novelty in both symbiotic partners. Although we have no traces or examples of the presumably long process that led to the eukaryotic cell, the *Tremblaya-Moranella* complex suggests that this process might have likely involved the modification or loss of old functions and the appearance of new capacities, reaching more or less viable intermediate stages until a global robust viability was reached. Viable internalisation could only have been achieved and maintained by increasing functional integration through new forms of global control, starting from the modulation of the permeability of the common boundary, to different systems of transport and targeting of functional components between the partners through the endosymbionts' membranes, which in turn made it possible to implement precise mechanisms of cross-control and interlocked regulation. All of them contributed towards the generation of a new and stronger form of individuality with a regulatory machinery in charge of all the internal functions, exemplified by the composite organisation of a eukaryotic cell, where the original endosymbiotic cells lost their former autonomy and became organelles. The emergence of a new functionally integrated organisation, therefore, requires a functional redefinition of both the original organisms and of the symbiotic consortium as a whole.

In sum, functional integration can generally be defined as the *degree* to which the different components of a biological dynamic regime of self-maintenance depend on one another for their *production*, *maintenance*, *activity*, and *reproduction*. If we take the eukaryotic cell as the reference example of new forms of full-fledged biological individuality resulting from association between prokaryotes, individuality can be understood in terms of the degree, scale and precision of the control and coordination of the parts that collectively make the system a viable functional whole (i.e. an integrated unit). To do so, even the minimal forms of biological (and, likely, proto-biological) organisation require, in the first place, some internal functional differentiation (Mossio et al., 2009). A cohesive integration between different functional tasks is achieved, then, when the differentiation of functions is coordinated at the system level by *control* and *regulatory mechanisms* that (1) act across the different entities participating in the association, and (2) are exerted in such a way that the differentiated components can contribute through their activity to the maintenance of the system. As we have seen, biological systems can give rise to different forms of functional associations, exhibiting different degrees of integration. In this process of integration, the deeper the co-dependency between the original organisations, the higher is their progressive loss of autonomy, accompanied by the appearance of new forms of control upon the members of the association.

References

Anderson, G. G., & O'Toole, G. A. (2008). Innate and induced resistance mechanisms of bacterial biofilms. In T. Romeo (Ed.), *Bacterial biofilms* (pp. 85–105). Springer. https://doi.org/10.1007/978-3-540-75418-3_5

Antunes, L. C. M., & Ferreira, R. B. R. (2009). Intercellular communication in bacteria. *Critical Reviews in Microbiology, 35*(2), 69–80. https://doi.org/10.1080/10408410902733946

Arnellos, A., Moreno, A., & Ruiz-Mirazo, K. (2014). Organizational requirements for multicellular autonomy: Insights from a comparative case study. *Biology & Philosophy, 29*(6), 851–884. https://doi.org/10.1007/s10539-013-9387-x

Bennett, G. M., & Moran, N. A. (2015). Heritable Symbiosis: The advantages and perils of an evolutionary rabbit hole. *Proceedings of the National Academy of Sciences of the United States of America, 112*(33), 10169–10176. https://doi.org/10.1073/pnas.1421388112

Bermejo, P., Sánchez, M. C., Llama-Palacios, A., Figuero, E., Herrera, D., & Sanz, M. (2019). Topographic characterization of multispecies biofilms growing on dental implant surfaces: An in vitro model. *Clinical Oral Implants Research, 30*(3), 229–241. https://doi.org/10.1111/clr.13409

Bich, L., Mossio, M., Ruiz-Mirazo, K., & Moreno, A. (2016). Biological regulation: Controlling the system from within. *Biology & Philosophy, 31*(2), 237–265. https://doi.org/10.1007/s10539-015-9497-8

Bich, L., Pradeu, T., & Moreau, J.-F. (2019). Understanding multicellularity: The functional organization of the intercellular space. *Frontiers in Physiology, 10*, 1170. https://doi.org/10.3389/fphys.2019.01170

Booth, A., & Ford Doolittle, W. (2015). Eukaryogenesis, how special really? *Proceedings of the National Academy of Sciences of the United States of America, 112*(33), 10278–10285. https://doi.org/10.1073/pnas.1421376112

Burmølle, M., Ren, D., Bjarnsholt, T., & Sørensen, S. J. (2014). Interactions in multispecies biofilms: Do they actually matter? *Trends in Microbiology, 22*(2), 84–91. https://doi.org/10.1016/j.tim.2013.12.004

Cairns, L. S., Hobley, L., & Stanley-Wall, N. R. (2014). Biofilm formation by bacillus subtilis: New insights into regulatory strategies and assembly mechanisms. *Molecular Microbiology, 93*(4), 587–598. https://doi.org/10.1111/mmi.12697

Catania, F., Krohs, U., Chittò, M., Ferro, D., Ferro, K., Lepennetier, G., Görtz, H.-D., Schreiber, R. S., Kurtz, J., & Gadau, J. (2017). The Hologenome concept: We need to incorporate function. *Theory in Biosciences, 136*(3), 89–98. https://doi.org/10.1007/s12064-016-0240-z

Cavalier-Smith, T. (2007). The Chimaeric origin of mitochondria: Photosynthetic cell enslavement, gene-transfer pressure, and compartmentation efficiency. In W. F. Martin & M. Müller (Eds.), *Origin of mitochondria and hydrogenosomes* (pp. 161–199). Springer. https://doi.org/10.1007/978-3-540-38502-8_8

Claessen, D., Rozen, D. E., Kuipers, O. P., Søgaard-Andersen, L., & van Wezel, G. P. (2014). Bacterial solutions to multicellularity: A tale of biofilms, filaments and fruiting bodies. *Nature Reviews. Microbiology, 12*(2), 115–124. https://doi.org/10.1038/nrmicro3178

Clarke, E. (2016). Levels of selection in biofilms: Multispecies biofilms are not evolutionary individuals. *Biology & Philosophy, 31*(2), 191–212. https://doi.org/10.1007/s10539-016-9517-3

Corsaro, D., & Venditti, D. (2006). Bacterial endosymbionts in prokaryotes. In J. M. Shively (Ed.), *Complex intracellular structures in prokaryotes* (pp. 359–371). Springer. https://doi.org/10.1007/7171_035

Dohlen, C. D. v., Shawn Kohler, S. T. A., & McManus, W. R. (2001). Mealybug Beta-proteobacterial endosymbionts contain gamma-proteobacterial symbionts. *Nature, 412*(6845), 433–436. https://doi.org/10.1038/35086563

Doolittle, W. F. (2013). Microbial Neopleomorphism. *Biology and Philosophy, 28*(2), 351–378. https://doi.org/10.1007/s10539-012-9358-7

Dragoš, A., & Kovács, Á. T. (2017). The peculiar functions of the bacterial extracellular matrix. *Trends in Microbiology, 25*(4), 257–266. https://doi.org/10.1016/j.tim.2016.12.010

Duncan, R. P., Husnik, F., Van Leuven, J. T., Gilbert, D. G., Dávalos, L. M., McCutcheon, J. P., & Wilson, A. C. C. (2014). Dynamic recruitment of amino acid transporters to the insect/symbiont interface. *Molecular Ecology, 23*(6), 1608–1623. https://doi.org/10.1111/mec.12627

Dupré, J., & O'Malley, M. (2009). Varieties of living things: Life at the intersection of lineage and metabolism. *Philosophy, Theory, and Practice in Biology, 1*(3), e003. https://doi.org/10.3998/ptb.6959004.0001.003

Elias, S., & Banin, E. (2012). Multi-species biofilms: Living with friendly neighbors. *FEMS Microbiology Reviews, 36*(5), 990–1004. https://doi.org/10.1111/j.1574-6976.2012.00325.x
Ereshefsky, M., & Pedroso, M. (2013). Biological individuality: The case of biofilms. *Biology & Philosophy, 28*(2), 331–349. https://doi.org/10.1007/s10539-012-9340-4
Ereshefsky, M., & Pedroso, M. (2015). Rethinking evolutionary individuality. *Proceedings of the National Academy of Sciences of the United States of America, 112*(33), 10126–10132. https://doi.org/10.1073/pnas.1421377112
Fidler, A. L., Darris, C. E., Chetyrkin, S. V., Pedchenko, V. K., Boudko, S. P., Brown, K. L., Gray Jerome, W., Hudson, J. K., Rokas, A., & Hudson, B. G. (2017). Collagen IV and basement membrane at the evolutionary dawn of metazoan tissues. *eLife, 6*, e24176. https://doi.org/10.7554/eLife.24176
Flemming, H.-C., Neu, T. R., & Wozniak, D. J. (2007). The EPS matrix: The "House of Biofilm Cells". *Journal of Bacteriology, 189*(22), 7945–7947. https://doi.org/10.1128/JB.00858-07
Flemming, H.-C., & Wingender, J. (2010). The biofilm matrix. *Nature Reviews Microbiology, 8*(9), 623–633. https://doi.org/10.1038/nrmicro2415
Griesemer, J. (2016). Reproduction in complex life cycles: Toward a developmental reaction norms perspective. *Philosophy of Science, 83*(5), 803–815. https://doi.org/10.1086/687865
Hammerschmidt, K., Rose, C. J., Kerr, B., & Rainey, P. B. (2014). Life cycles, fitness decoupling and the evolution of multicellularity. *Nature, 515*(7525), 75–79. https://doi.org/10.1038/nature13884
Howes, M. (1998). The self of philosophy and the self of immunology. *Perspectives in Biology and Medicine, 42*(1), 118–130. https://doi.org/10.1353/pbm.1998.0046
Hull, D. L. (1980). Individuality and selection. *Annual Review of Ecology, Evolution, and Systematics, 11*, 311–332. https://doi.org/10.1146/annurev.es.11.110180.001523
Julien, B., Dale Kaiser, A., & Garza, A. (2000). Spatial control of cell differentiation in Myxococcus Xanthus. *Proceedings of the National Academy of Sciences of the United States of America, 97*(16), 9098–9103. https://doi.org/10.1073/pnas.97.16.9098
Karatan, E., & Watnick, P. (2009). Signals, regulatory networks, and materials that build and break bacterial biofilms. *Microbiology and Molecular Biology Reviews, 73*(2), 310–347. https://doi.org/10.1128/MMBR.00041-08
Keller, L., & Surette, M. G. (2006). Communication in bacteria: An ecological and evolutionary perspective. *Nature Reviews Microbiology, 4*(4), 249–258. https://doi.org/10.1038/nrmicro1383
Lane, N. (2005). *Power, sex, suicide*. Oxford University Press.
Lane, N. (2015). *The vital question: Energy, evolution, and the origins of complex life*. Norton & Company.
Lill, R. (2009). Function and biogenesis of iron-sulphur proteins. *Nature, 460*(7257), 831–838. https://doi.org/10.1038/nature08301
Liu, W., Røder, H. L., Madsen, J. S., Bjarnsholt, T., Sørensen, S. J., & Burmølle, M. (2016). Interspecific bacterial interactions are reflected in multispecies biofilm spatial organization. *Frontiers in Microbiology, 7*, 1366. https://doi.org/10.3389/fmicb.2016.01366
Lohse, M. B., Gulati, M., Johnson, A. D., & Nobile, C. J. (2018). Development and regulation of single- and multi-species Candida Albicans biofilms. *Nature Reviews Microbiology, 16*(1), 19–31. https://doi.org/10.1038/nrmicro.2017.107
Lopez,D.,Vlamakis,H.,&Kolter,R.(2009).Generationofmultiplecelltypesinbacillussubtilis.*FEMS Microbiology Reviews, 33*(1), 152–163. https://doi.org/10.1111/j.1574-6976.2008.00148.x
López-Madrigal, S., Latorre, A., Porcar, M., Moya, A., & Gil, R. (2011). Complete genome sequence of "Candidatus Tremblaya Princeps" strain PCVAL, an intriguing translational machine below the living-cell status. *Journal of Bacteriology, 193*(19), 5587–5588. https://doi.org/10.1128/JB.05749-11
López-Madrigal, S., Latorre, A., Porcar, M., Moya, A., & Gil, R. (2013b). Mealybugs nested endosymbiosis: Going into the "Matryoshka" system in Planococcus Citri in depth. *BMC Microbiology, 13*(1), 74. https://doi.org/10.1186/1471-2180-13-74

López-Madrigal, S., Balmand, S., Latorre, A., Heddi, A., Moya, A., & Gil, R. (2013a). How does Tremblaya Princeps get essential proteins from its nested partner Moranella Endobia in the mealybug Planoccocus Citri? *PLoS One, 8*(10), e77307. https://doi.org/10.1371/journal.pone.0077307

Lyons, N. A., & Kolter, R. (2015). On the evolution of bacterial multicellularity. *Current Opinion in Microbiology, 24*, 21–28. https://doi.org/10.1016/j.mib.2014.12.007

Martin, W. (2010). Evolutionary origins of metabolic compartmentalization in eukaryotes. *Philosophical Transactions of the Royal Society B: Biological Sciences, 365*(1541), 847–855. https://doi.org/10.1098/rstb.2009.0252

McCutcheon, J. P., & von Dohlen, C. D. (2011). An interdependent metabolic patchwork in the nested symbiosis of mealybugs. *Current Biology, 21*(16), 1366–1372. https://doi.org/10.1016/j.cub.2011.06.051

Militello, G., Bich, L., & Moreno, A. (2021). Functional integration and individuality in prokaryotic collective organisations. *Acta Biotheoretica, 69*(3), 391–415. https://doi.org/10.1007/s10441-020-09390-z

Moreno, A., & Mossio, M. (2015). *Biological autonomy: A philosophical and theoretical enquiry*. Springer.

Morris, J. J., Lenski, R. E., & Zinser, E. R. (2012). The black queen hypothesis: Evolution of dependencies through adaptive gene loss. *mBio, 3*(2), e00036–e00012. https://doi.org/10.1128/mBio.00036-12

Mossio, M., Saborido, C., & Moreno, A. (2009). An organizational account of biological functions. *British Journal for the Philosophy of Science, 60*(4), 813–841. https://doi.org/10.1093/bjps/axp036

Muñoz-Dorado, J., Marcos-Torres, F. J., García-Bravo, E., Moraleda-Muñoz, A., & Pérez, J. (2016). Myxobacteria: Moving, killing, feeding, and surviving together. *Frontiers in Microbiology, 7*, 781. https://doi.org/10.3389/fmicb.2016.00781

Pradeu, T. (2010). What is an organism? An immunological answer. *History and Philosophy of the Life Sciences, 32*(2–3), 247–267.

Queller, D. C., & Strassmann, J. E. (2009). Beyond society: The evolution of organismality. *Philosophical Transactions of the Royal Society of London Series B, Biological Sciences, 364*(1533), 3143–3155. https://doi.org/10.1098/rstb.2009.0095

Queller, D. C., & Strassmann, J. E. (2016). Problems of multi-species organisms: Endosymbionts to holobionts. *Biology & Philosophy, 31*(6), 855–873. https://doi.org/10.1007/s10539-016-9547-x

Rainey, P. B., & Kerr, B. (2010). Cheats as first propagules: A new hypothesis for the evolution of individuality during the transition from single cells to multicellularity. *BioEssays: News and Reviews in Molecular, Cellular and Developmental Biology, 32*(10), 872–880. https://doi.org/10.1002/bies.201000039

Reyes-Prieto, M., Latorre, A., & Moya, A. (2014). Scanty microbes, the "symbionelle" concept. *Environmental Microbiology, 16*(2), 335–338. https://doi.org/10.1111/1462-2920.12220

Saxena, P., Joshi, Y., Rawat, K., & Bisht, R. (2019). Biofilms: Architecture, resistance, quorum sensing and control mechanisms. *Indian Journal of Microbiology, 59*(1), 3–12. https://doi.org/10.1007/s12088-018-0757-6

Shapiro, J. A., Dworkin, M., Shapiro, J. A., & Dworkin, M. (Eds.). (1997). *Bacteria as multicellular organisms*. Oxford University Press.

Skillings, D. (2016). Holobionts and the ecology of organisms: Multi-species communities or integrated individuals? *Biology & Philosophy, 31*(6), 875–892. https://doi.org/10.1007/s10539-016-9544-0

Skillings, D. (2019). Trojan horses and black Queens: Causal Core? Explanations in microbiome research. *Biology and Philosophy, 34*(6), 1–6. https://doi.org/10.1007/s10539-019-9718-7

Sober, E. (1991). Organisms, individuals, and units of selection. In A. I. Tauber (Ed.), *Organism and the origins of self. Boston studies in the philosophy of science* (Vol. 129, pp. 275–296). Springer. https://doi.org/10.1007/978-94-011-3406-4_13

Solano, C., Echeverz, M., & Lasa, I. (2014). Biofilm dispersion and quorum sensing. *Current Opinion in Microbiology, 18*, 96–104. https://doi.org/10.1016/j.mib.2014.02.008

Steinberg, N., & Kolodkin-Gal, I. (2015). The matrix reloaded: Probing the extracellular matrix synchronizes bacterial communities. *Journal of Bacteriology, 197*(13), 2092–2103. https://doi.org/10.1128/JB.02516-14

Stewart, P. S., & William Costerton, J. (2001). Antibiotic resistance of bacteria in biofilms. *Lancet, 358*(9276), 135–138. https://doi.org/10.1016/s0140-6736(01)05321-1

Sutherland, I. W. (2001). The biofilm matrix—an immobilized but dynamic microbial environment. *Trends in Microbiology, 9*(5), 222–227. https://doi.org/10.1016/s0966-842x(01)02012-1

Wilson, E. O. (1974). The perfect societies: Animal colonies. Development and function through time. *Science, 184*(4132), 54–55. https://doi.org/10.1126/science.184.4132.54

Wolcott, R., William Costerton, J., Raoult, D., & Cutler, S. J. (2013). The polymicrobial nature of biofilm infection. *Clinical Microbiology and Infection: The Official Publication of the European Society of Clinical Microbiology and Infectious Diseases, 19*(2), 107–112. https://doi.org/10.1111/j.1469-0691.2012.04001.x

Wujek, D. E. (1979). Intracellular bacteria in the blue-green alga Pleurocapsa minor. *Transactions of the American Microscopical Society, 98*(1), 143–145. https://doi.org/10.2307/3225953

Chapter 6
Divide Et Impera: How Cellular Integration and Control Are Established Through the Division of the Intracellular Space

Abstract This chapter explores how eukaryotic cells achieved new levels of functional integration through the division of their internal space into intracellular membrane-bound compartments. By comparing the internal membranes of prokaryotes and eukaryotes, it argues that these organelles partition the intracellular space in different ways and perform distinct functions, enhancing overall cellular operations. In prokaryotes, constitutive processes primarily occur in the cytoplasm, which exerts near-total control over the microcompartments. In eukaryotes, however, control over these processes is distributed between the cytoplasm and organelles, leading to a more balanced regulation of the cell's global physiology. These varying modes of control create distinct forms of integration between the cytoplasm and microcompartments.

Keywords Intracellular membranes · Prokaryotes · Eukaryotes · Functional coordination · Functional integration

6.1 Introduction

For a long time, it was believed that intracellular membrane-bound compartments were a unique feature of eukaryotic cells. However, some species of bacteria and archaea also possess these compartments (Shively, 2006; Murat et al., 2010). Both prokaryotic and eukaryotic microcompartments play important roles in cellular physiological processes such as metabolism, genetic functions, and the cell cycle. Despite this, key differences exist between prokaryotic and eukaryotic organelles in terms of morphology and functionality. Eukaryotic organelles exhibit a greater variety of structures and functions compared to their prokaryotic counterparts.

While Chap. 5 explored how a common physical boundary surrounding prokaryotic collective organisations enables a certain type and degree of functional integration, this chapter examines the role of prokaryotic and eukaryotic membrane-bound compartments in the internal division of the cytoplasmic space and their implications for cellular physiological integration. This topic will be addressed by studying

G. Militello, *Functional Integration*, History, Philosophy and Theory of the Life Sciences 38, https://doi.org/10.1007/978-3-031-94744-5_6

the main physiological mechanisms that enable bacterial and archaeal microcompartments, as well as eukaryotic endomembranes,[1] to integrate functionally within the cellular network. Additionally, the evolutionary origins and potential evolutionary opportunities presented by these intracellular structures will be discussed.

Thus, this chapter examines the following research questions: What are the organisational features of prokaryotic and eukaryotic endomembranes and organelles? What types and degrees of functional integration are enabled by prokaryotic and eukaryotic microcompartments? What is the origin of, and what evolutionary possibilities are opened up by, prokaryotic and eukaryotic membrane-bound compartments?

It will be argued that prokaryotic and eukaryotic organelles divide the intracellular space differently and perform distinct functions. In both cases, this internal division enhances overall cellular functions. This results in distinct forms of control over the cell's constitutive processes. In prokaryotes, these processes primarily occur in the cytoplasm, which exerts almost complete control over the microcompartments. In contrast, in eukaryotes, control over constitutive processes is shared between the cytoplasm and the organelles, leading to a more uniform distribution of control over the cell's global physiology. These differing modes of control result in distinct forms of integration between the cytoplasm and microcompartments.

The chapter is structured as follows. Section 6.2 reviews the main organisational properties of biological membranes. Sections 6.3, 6.4, and 6.5 analyse the organisational features of prokaryotic microcompartments (Sect. 6.3) and eukaryotic endomembranes (Sects. 6.4 and 6.5). Section 6.6 compares prokaryotic and eukaryotic organelles and examines their different contributions to the physiological integration of cells. Finally, Sect. 6.7 provides concluding remarks.

6.2 A Critical View on (Intra)Cellular Membranes

Biological membranes play a pivotal role in cellular organisation due to their physicochemical properties, which are essential for maintaining the internal chemical network and, more broadly, cellular life. Significant conceptual effort has been devoted to understanding the organisational properties of the plasma membrane, both in biology (Edidin, 2003; Goñi, 2014; Bernardino de la Serna et al., 2016) and in the philosophy of biology(Maturana & Varela, 1973; Varela, 1979; Moreno & Mossio, 2015). However, biological membranes encompass not only the plasma membrane but also the internal membranes that divide the cytoplasmic space into

[1] This chapter examines the nuclear envelope and the membranes of organelles within the endomembrane system as examples of eukaryotic endomembranes. The membranes of organelles with endosymbiotic origins (mitochondria, plastids, and hydrogenosomes) are excluded for two main reasons: first, their role was addressed conceptually in Chap. 5. Second, the analysis of the nucleus and endomembrane system suffices to support my argument, avoiding the need for a further section on the membranes of endosymbiotic organelles.

subregions, found in both eukaryotic cells and several prokaryotes (bacteria and archaea). This chapter reviews the main biological properties of cellular membranes to characterise the organisational features of intracellular membranes and examines how these characteristics apply to membrane-bound compartments.

A fundamental tenet of the autopoietic and organisational framework is that a physical boundary permits the distinction between the *interior* and *exterior* of the cell, establishing a clear separation between the living system and its surroundings (Maturana & Varela, 1973; Varela, 1979; Moreno & Mossio, 2015). In the case of intracellular membranes, this aspect is more complex because they do not separate the inside from the outside of the cell. Nevertheless, to a certain extent, the analogy with the cell membrane holds true. Internal membranes divide the inner cellular space, creating specific *subregions* that allow for the coexistence of different, sometimes incompatible, biochemical pathways and enhance the efficiency of metabolic processes, thereby improving overall cell physiology (Helle et al., 2013; Gabaldón & Pittis, 2015). Furthermore, the membranes of some eukaryotic organelles, such as the endoplasmic reticulum and the Golgi apparatus, increase the surface area and volume of these organelles. This expansion allows a larger number of molecules to be accommodated within the membranes, thereby increasing metabolic fluxes (Marshall, 2012).

Since membranes act as interfaces between different environments, they can undergo remodelling—changes in membrane composition and shape—in response to environmental cues. These changes, such as replacing loosely packed unsaturated lipids with tightly packed saturated ones, aim to preserve the physicochemical properties of membranes (e.g. viscosity, surface charge density, thickness), which must be homeostatically maintained within a narrow range compatible with cellular physiology (Ernst et al., 2018). Many cells, ranging from prokaryotes to those of ectothermic animals, adapt their membrane properties to changes in ambient temperature to maintain membrane physical properties. For example, *Bacillus subtilis* can adjust the thickness of its plasma membrane in response to cold temperatures (Cybulski et al., 2010). The membranes of organelles also undergo remodelling in response to internal metabolic perturbations. For instance, the membrane of the endoplasmic reticulum can be remodelled due to changes in lipid synthesis (Ernst et al., 2018). This plasticity of biological membranes allows cells and membrane-bound compartments to respond adequately to external and internal variations, representing a fundamental contribution to cell physiology.

Cell membranes function as spatial constraints that selectively regulate the flow of molecules and metabolites into and out of the cell. Molecules can pass through the membrane either passively or actively,[2] with their movement mediated by transmembrane proteins such as channels and transporters. Similarly, both prokaryotic and eukaryotic endomembranes serve as spatial constraints by acting as selective

[2] In passive transport, molecules move by exploiting the concentration differences across the membrane through simple diffusion or transmembrane proteins. In active transport, molecules move against the concentration gradient, requiring additional energy from sources such as ATP hydrolysis to overcome the energetic barrier.

barriers that enable both passive and active transport of substances to and from organelles. For example, bacterial microcompartments feature protein shells with pores that act as selective permeability barriers. These pores control the movement of enzyme cofactors, substrates, and products between the microcompartment and the cytoplasm through both passive diffusion and dynamically gated mechanisms (Chowdhury et al., 2014; Bobik et al., 2015). In eukaryotic cells, membrane-bound compartments also contain transmembrane proteins that regulate the transport of macromolecules across their membranes. For instance, nuclear pores in the nuclear envelope act as selective barriers between the nucleoplasm and the cytoplasm (Lamond & Sleeman, 2003).

Cell membranes are functionally interdependent with metabolism: membranes constrain the flow of (macro)molecules to maintain appropriate substrate concentrations within the metabolic network, while metabolism sustains the physical boundary of membranes by synthesising their constitutive components, such as lipids, proteins, glycoproteins, and glycolipids (Maturana & Varela, 1973; Moreno & Mossio, 2015). This relationship raises questions about how intracellular membranes and cellular metabolism interact: if organelles perform metabolic functions, are they involved in their own membrane biogenesis, or do they rely on metabolic reactions occurring in the cytoplasm?

The answer depends on where lipids, proteins, glycoproteins, and glycolipids are synthesised. For prokaryotic microcompartments, these macromolecules are produced in the cytoplasm, and internal membranes likely form through self-assembly and invagination mechanisms involving cytoplasmic proteins (Murat et al., 2010). In contrast, eukaryotic internal membranes present a more complex scenario. Proteins are synthesised both in cytoplasmic ribosomes and in ribosomes of the rough endoplasmic reticulum. Most membrane phospholipids are produced in the Golgi complex, while others, such as cholesterol, are synthesised in the endoplasmic reticulum (Simons & Sampaio, 2011). Consequently, the biogenesis of internal membranes is intricately linked with metabolic processes occurring both in the cytoplasm and within organelles, necessitating new forms of intracellular transport, such as vesicles.

Cell membranes facilitate interactions between cells and their surroundings through the exchange of signalling molecules. They contain numerous transmembrane receptors that detect extracellular cues and signals—such as nutrients, prey, or signals from other cells—and initiate signalling cascades that can activate or inhibit various cellular processes. These receptors also enable cells to send signals to other cells, promoting intercellular communication.

Intracellular membranes have their own set of receptors involved in intracellular signalling pathways, which coordinate functions between organelles and the cytoplasm, as well as among different organelles. For example, the gated pores of prokaryotic intracellular membranes exchange signals with cytoplasmic membranes, coordinating metabolic reactions occurring within the microcompartments and the cytoplasm (Chowdhury et al., 2014). In eukaryotic cells, intracellular membranes are part of a complex system of inter-organelle communication, which ensures functional coordination among various organelles (Hieda, 2019).

As discussed in Chap. 5, physical boundaries, including cytoplasmic membranes, contribute to the physiological integration of cellular organisation by constraining and coordinating various physiological mechanisms—such as metabolic, signalling, and immunologic processes. If this analogy holds true for intracellular membranes, we can hypothesise that these membranes also play a role in the functional integration of the cell. Specifically, they may facilitate functional integration between the organelles they enclose and the surrounding cytoplasmic space.

Given that intracellular membranes are present in both prokaryotes and eukaryotes, it is pertinent to investigate whether these membranes enable similar types of physiological integration or if there are organisational differences between prokaryotic and eukaryotic systems. To explore these issues, I will examine in the following sections the key physiological mechanisms involved in the functional integration of membrane-bound compartments with the cytoplasm.

6.3 Prokaryotic Microcompartments

For several decades, bacteria and archaea were thought to lack internal compartments, with the endomembrane system considered a feature unique to eukaryotic cells. However, research has revealed that many bacteria (e.g. cyanobacteria, magnetotactic bacteria, planctomycetes) and some archaea (e.g. *Ignicoccus hospitalis*) possess protein- or lipid-bound compartments[3] (Shively, 2006; Murat et al., 2010; Diekmann & Pereira-Leal, 2013) (Fig. 6.1). These compartments create specialised intracellular metabolic niches. This section will explore the organisational features of these protein- and lipid-bound microcompartments and examine the mechanisms by which they are integrated within prokaryotic cells.

Two types of organelles are encased by *proteinaceous* membranes: *carboxysomes* and *gas vesicles*.

1. *Carboxysomes* (Fig. 6.1.1a): Found in cyanobacteria and chemoautolithotrophic bacteria, carboxysomes are the sites where the initial steps of the Calvin cycle occur. These organelles contain the enzyme RuBisCo[4] and carbonic anhydrase,[5] which together enhance the efficiency of carbon fixation (Yeates et al., 2008; Kerfeld et al., 2018) (see Fig. 6.1).
2. *Gas Vesicles* (Fig. 6.1.1b): Present in certain bacteria and archaea, particularly those in aquatic environments, gas vesicles regulate buoyancy. Their hydrophobic nature prevents water molecules from forming droplets on the inner surface,

[3] These organelles originate from the inward folding (tubulation, invagination, or vesiculation) of the prokaryotic plasma membrane, giving them an autogenous rather than endosymbiotic origin.

[4] Ribulose-1,5-bisphosphate carboxylase/oxygenase (RuBisCo) is an enzyme involved in a critical carbon fixation pathway, catalysing the reaction of CO_2 with ribulose bisphosphate to form two molecules of 3-phosphoglyceric acid.

[5] Carbonic anhydrase catalyses the conversion of bicarbonate into CO_2 (Murat et al., 2010).

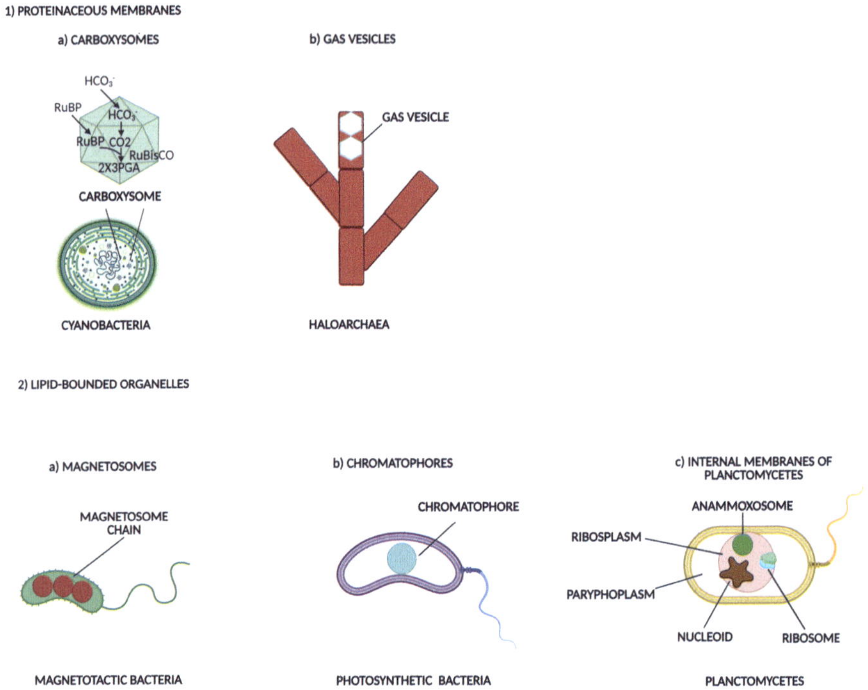

Fig. 6.1 Diverse prokaryotic microcompartments. (1) Proteinaceous membranes: (a) Carboxysomes in cyanobacteria, crucial for carbon fixation through the encapsulation of RuBisCO; (b) Gas vesicles in Haloarchaea, providing buoyancy control. (2) Lipid-bounded organelles: (a) Magnetosomes in magnetotactic bacteria, facilitating geomagnetic orientation; (b) Chromatophores in photosynthetic bacteria, involved in light energy capture; (c) Internal membranes of planctomycetes, illustrating cellular compartmentalisation for specialised metabolic functions

allowing these vesicles to control the vertical movement of the cells within the water column. This buoyancy regulation also affects the cells' exposure to light, salinity, and other environmental factors (Pfeifer, 2015).

Lipid-bound organelles can be classified into three main categories:

1. *Magnetosomes* (Fig. 6.1.2a): Found in magnetotactic bacteria, magnetosomes are spherical lipid-bound compartments that enclose magnetic particles. These organelles help bacteria align with the Earth's magnetic field,[6] facilitating their navigation (Grant et al., 2018). The magnetosome membrane, which is continuous with or derived from the inner cell membrane, contains specific proteins and is supported by a network of actin-like filaments that regulate magnetosome positioning (Scheffel et al., 2006; Grant et al., 2018).

[6]Although magnetosomes do not directly contribute to metabolism, they indirectly affect it by enabling magnetotactic bacteria to navigate for oxygen.

2. *Photosynthetic Membranes*[7] (Fig. 6.1.2b): Present in photosynthetic bacteria, such as cyanobacteria, these membranes include chromatophores and chlorosomes. They are crucial for optimising the photosynthetic process by providing large surface areas for light absorption and reaction (Pfeifer, 2015).
3. *Internal Membranes of Planctomycetes* (Fig. 6.1.2c): Planctomycetes feature unique internal membranes that are distinct from the conventional cellular membranes. These membranes are involved in various cellular processes, contributing to the cellular organisation of these bacteria.

Planctomycetes represent an intriguing case of bacterial organelles with distinct compartmentalisation. Their cytoplasm is divided into two main compartments: the *riboplasm*, which contains the nucleoid and ribosomes, and the *paryphoplasm*, situated between the outer and inner membranes and typically lacking ribosomes but often containing vesicles.[8] Unlike typical Gram-negative bacteria, planctomycetes feature a highly invaginated outer membrane, which resembles mitochondrial membranes (Santarella-Mellwig et al., 2013). Additionally, some planctomycetes, such as *Gemmata obscuriglobus*, have an extra double-membraned compartment that houses the nucleoid (Lindsay et al., 2001). They also possess an *anammoxosome*, an organelle enclosed by a single lipid bilayer, which is crucial for anaerobic ammonium oxidation (anammox). The anammoxosome's membrane serves as a diffusion barrier, conserving energy by minimising passive proton diffusion during slow anammox metabolism and retaining reaction intermediates (Neumann et al., 2014; Grant et al., 2018). This membrane is also highly invaginated, likely to enhance energy efficiency, and a few enzymes involved in ammonium oxidation are specifically targeted to the anammoxosome matrix through signal peptides for the Sec or Tat translocation systems (de Almeida et al., 2015; Grant et al., 2018).

Membrane-bound compartments are crucial for prokaryotic physiology, significantly enhancing specific *metabolic pathways* by encapsulating and concentrating enzymes and intermediates. For instance, photosynthetic organelles increase the efficiency of photosynthesis by expanding the membrane surface area exposed to light and increasing the number of photosynthetic protein complexes (Murat et al., 2010). Similarly, prokaryotic organelles play a key role in isolating toxic or volatile metabolic intermediates from the rest of the cytoplasm, preventing their diffusion and potential disruption. For example, carboxysomes sequester toxic aldehydes like propionaldehyde and acetaldehyde, while also containing CO_2, thus preventing its rapid diffusion across the cell envelope (Chowdhury et al., 2014; Kirst & Kerfeld,

[7] Photosynthetic membranes are categorised into three main groups: chromatophores, thylakoids, and chlorosomes (Murat et al., 2010). Chromatophores form from protein assemblies in the inner cell membrane that invaginate. Thylakoids consist of layered structures that form a circle. Chlorosomes likely result from a self-assembly process of lipids and proteins, distinct from the mechanisms forming other lipid-bound organelles.

[8] Based on their morphological features, planctomycetes have been proposed as an intermediate step between prokaryotic and eukaryotic cells. However, this hypothesis has been criticised by McInerney et al., (2011), who argue that planctomycetes represent an analogy, not homology, to eukaryotic organelles.

2019). Additionally, the anammoxosomes of planctomycetes retain intermediates of slow anammox reactions, enhancing energy conservation by reducing passive proton diffusion (Chowdhury et al., 2015). Some organelles, such as magnetosomes, influence global cellular behaviour by orienting magnetotactic bacteria within an external magnetic field via a network of actin-like filaments (Murat et al., 2010).

The functions of prokaryotic microcompartments are deeply intertwined with both bacterial and archaeal metabolism and the genetic and metabolic processes in the cytoplasm. The protein components of these microcompartments are synthesised and regulated by the organism's genome, with genes often clustered in operons within the cytoplasm. For instance, the proteins forming the membranes of carboxysomes and gas vesicles are encoded by several genes grouped together (Murat et al., 2010). Additionally, the efficiency of metabolic processes within these microcompartments relies on the availability of metabolites, which must be transported from the cytoplasm. Although the mechanisms are not fully elucidated, it is hypothesised that metabolites enter microcompartments via concentration gradients (Bobik et al., 2015), gated pores that respond to cytoplasmic signals reflecting the cell's metabolic state (Chowdhury et al., 2014; Bobik et al., 2015), or through the colocalisation of sequentially acting enzymes, which facilitates the diffusion of metabolic intermediates across the membrane (Bobik et al., 2015).

The *positioning* of prokaryotic organelles is not random; it is carefully orchestrated by cytoskeletal-like elements and motor proteins[9] that provide structural organisation and dynamic movement within the cell (Savage et al., 2010). For example, bacterial cytoskeleton-like proteins such as MreB and ParA utilise the energy from ATP or GTP hydrolysis to generate mechanical forces that arrange carboxysomes linearly within the cell, ensuring their functional integration with other cellular components and even distribution during cell division (Savage et al., 2010). Similarly, magnetosomes align into chains due to the action of actin-like proteins like MamB, MamM, and MamE, which polymerise to form a protein complex that exerts lateral pressure, inducing membrane curvature (Raschdorf et al., 2016; Grant et al., 2018). These cytoskeletal-like proteins also play a critical role in the segregation of organelles; for instance, carboxysomes and magnetosomes are segregated by bacterial cytoskeletal proteins such as ParA and MamK, respectively (Cornejo et al., 2014; Toro-Nahuelpan et al., 2016).

In light of the above discussion, two key conclusions emerge. First, the primary function of prokaryotic microcompartments is to isolate and *enhance metabolic reactions*, thereby optimising cellular metabolism. Second, both the structural components and functionality of these microcompartments are intricately linked to the *genetic and metabolic processes* occurring in the cytoplasm. This physiological integration is achieved through various mechanisms, including the movement of metabolites between the cytoplasm and the microcompartment lumen, as well as the association of microcompartments with cytoskeletal proteins.

---

[9] For a deeper discussion of motor proteins, see Militello and Moreno (2018).

6.4 The Nuclear Envelope and the Compartmentalisation of the Genetic Material

The nuclear envelope (NE) is a highly specialised membrane structure composed of two phospholipid bilayers: the inner nuclear membrane (INM), which faces the nucleoplasm, and the outer nuclear membrane (ONM), which is adjacent to the cytoplasm and continuous with the rough endoplasmic reticulum (rER) (Fig. 6.2). These membranes are perforated by nuclear pore complexes (NPCs), which are intricate multiprotein assemblies made up of numerous nucleoporins. The inner surface of the NE is lined by the nuclear lamina,[10] a network of filamentous proteins that provides structural support. While the NE serves to spatially *separate* the nucleoplasm from the cytoplasm, it also maintains structural and functional *connectivity* with both the cytoplasm and the endoplasmic reticulum. This dual role

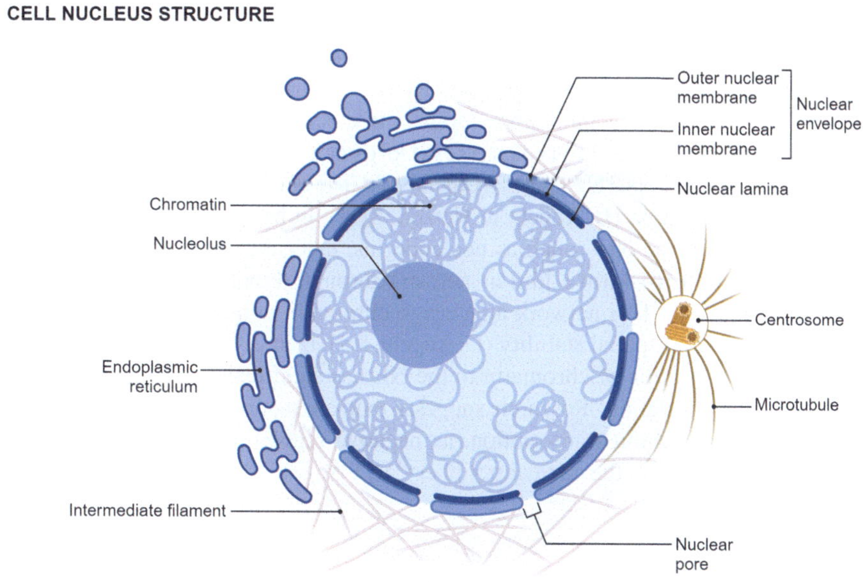

Fig. 6.2 Detailed structure of the eukaryotic nucleus and nuclear envelope. This figure illustrates the organisation of the eukaryotic nucleus, including the double-membraned nuclear envelope, nuclear pores, chromatin, nucleolus, and the underlying nuclear lamina. The nuclear envelope regulates the exchange of materials between the nucleus and cytoplasm, while the nucleolus is the site of ribosomal RNA synthesis. Chromatin, composed of DNA and histone proteins, is organised into euchromatin and heterochromatin, reflecting regions of active and inactive transcription, respectively

[10] Although the nuclear lamina is widespread among eukaryotes, some protozoa (e.g. *Saccharomyces cerevisiae*) lack it but have proteins (e.g. Mplp1 and Esc1) that functionally replace the lamina (Rout et al., 2017).

facilitates the seamless *integration* of the nucleoplasm with the cytoplasm and other eukaryotic subregions. This section will focus on the structural and functional organisation of the nuclear envelope and its role in the functional integration of the eukaryotic cell.

The nuclear envelope (NE) provides a physical *separation* between the cytoplasm and the nucleoplasm, which contains several membraneless compartments, such as nucleoli,[11] Cajal bodies, and splicing speckles (Lamond & Sleeman, 2003). These compartments play crucial roles not only in genetic transcription but also in various metabolic functions, such as modulating the access of nuclear enzymes and receptors to their substrates (Lamond & Sleeman, 2003). One of the significant evolutionary advancements afforded by the NE is the spatial and temporal separation of transcription and translation processes: transcription occurs within the nucleus, while translation takes place in the cytoplasm or the rough endoplasmic reticulum. This separation has allowed eukaryotic cells to develop a more sophisticated regulatory framework compared to prokaryotic cells. For instance, eukaryotic cells can engage in splicing of primary transcripts[12] before translation, selectively regulate transcriptional access to chromatin, and employ new epigenetic mechanisms, such as histone modifications and RNA interference. Similarly, the regulation of translation has become more complex in eukaryotes, with a wider array of initiation factors compared to prokaryotes (Devos et al., 2014).

The nuclear envelope (NE) plays a crucial role in *organising chromatin* and *influencing gene expression* (D'Angelo & Hetzer, 2006). In both protozoans and metazoans, proteins of the NE—such as nucleoporins, inner nuclear membrane (INM) proteins, and lamins—form a fibrous network through direct and indirect protein-protein interactions. This network attaches to chromatin domains, providing structural anchorage and spatial stability (D'Angelo & Hetzer, 2006; Gavrilov & Razin, 2015). The localisation of chromatin to the NE has significant functional implications for gene regulation. Research suggests that interactions between chromatin and the NE may inhibit transcription, as negative regulators of transcription are often found localised at the nuclear periphery[13] (D'Angelo & Hetzer, 2006; Padeken & Heun, 2014).

---

[11] Nucleoli are involved in rRNA synthesis and ribosomal unit assembly, exhibiting a dynamic structure that cycles through assembly and disassembly during the cell cycle. Cajal bodies, which do not contain rRNA or rRNA genes, are mainly involved in the maturation of nuclear ribonucleoprotein complexes, including snRNPs and snoRNPs. Splicing speckles act as reservoirs for splicing factors of nascent pre-mRNA at nearby genes. The functions of other nuclear bodies (e.g. PML bodies, clastosomes, paraspeckles) are not clearly understood (Lamond & Sleeman, 2003).

[12] This post-transcriptional regulation is physiologically crucial because the translation of unspliced pre-mRNA would produce non-functional proteins that could also negatively inhibit translation (Görlich & Kutay, 1999).

[13] This applies to both protozoa and metazoans. For example, in the protozoan *Trypanosoma brucei*, nucleoporins such as NUP1 are involved in gene silencing, including the silencing of variant surface glycoprotein by NUP1 (Obado et al., 2016). Similarly, the nuclear envelope protein MAN1 inhibits bone morphogenic protein signalling during Xenopus development (Osada et al., 2003).

The nuclear envelope (NE) plays a crucial role not only in separating the nucleoplasm from the cytoplasm but also in regulating the *selective passage* of macromolecules between these compartments and the rough endoplasmic reticulum (rER). RNA molecules, such as tRNA, mRNA, and rRNA, are synthesised in the nucleus and must be exported to the cytoplasm or rER for translation and processing. Conversely, proteins required for nuclear functions—such as histones, transcription factors, and splicing factors—are synthesised in the cytoplasm or rER and need to be imported into the nucleus (Lamond & Sleeman, 2003). This transport is mediated by nuclear pore complexes (NPCs) and involves various receptors like importins and exportins.[14] For nuclear import, the importin α/β heterodimer binds to protein cargos and the RanGTP complex, and importin β facilitates interaction with NPCs, allowing the release of cargo within the nucleus (Görlich & Kutay, 1999). Proteins synthesised in the rER can pass through the nuclear envelope via either rapid passive diffusion (for small proteins) or active transport through NPCs (for larger proteins) (D'Angelo & Hetzer, 2006). In nuclear export, RNA molecules bind to mobile export receptors—small RNAs (e.g. tRNAs, microRNAs) directly to exportins, and large RNAs (e.g. mRNAs, rRNAs) through specific adaptor proteins after assembling into ribonucleoprotein particles (Köhler & Hurt, 2007).

A major organisational feature of the nucleus is its interaction with the *cytoskeleton*, which includes microtubules, actin filaments, and intermediate filaments. This interaction is mediated by proteins such as Nesprins, predominantly located in the outer nuclear membrane (ONM), and the LINC complex (Linker of Nucleoskeleton and Cytoskeleton), which includes SUN-domain proteins and KASH proteins[15] that span both membranes of the nuclear envelope (NE). Nesprins and KASH proteins bind to cytoskeletal motor proteins, leveraging energy to generate mechanical forces that enable the displacement of the nucleus along cytoskeletal filaments. This coupling between the NE and the cytoskeleton is crucial for the functional integration of the nucleus within the eukaryotic cell network and has several important physiological consequences.

First, it facilitates the *positioning and movement of the nucleus* within the cell, influencing cellular architecture and function. Nesprins, bound to the domains of the LINC complex, connect the nuclear envelope (NE) to microtubules and actin filaments, thereby regulating nuclear positioning and movement (Tapley & Starr, 2013). This connection allows for two primary mechanisms of nuclear movement: the recruitment of motor proteins such as dynein and kinesin, which generate force to move the nucleus along microtubules, and the tethering of the nucleus to a dynamic actin network, a process also facilitated by SUN proteins. The positioning and movement of the nucleus are thought to influence the organisation and mechanical properties of the cytoskeleton, cytoplasmic signalling, and the accessibility of the nucleus to various signalling pathways (Gundersen & Worman, 2013). The NE

[14] Importins and exportins are proteins that mediate the import and export of macromolecules, respectively.

[15] SUN-domain proteins directly interact with KASH proteins.

acts as a cytoskeletal integrator, contributing to the organisation of microtubules by binding to molecules such as microtubule-organising centres (MTOCs)[16] and interacting with actin through nesprins (Gundersen & Worman, 2013). Additionally, nuclear movement generates mechanical forces that can modulate cytoplasmic signalling pathways, including Rho GTPase and MAP kinase pathways, and affect the nucleus's responsiveness to signals that regulate transcription and mRNA transport. The spatial positioning of the nucleus may also influence the access of transcription factors and second messengers from the cytoplasm into the nucleoplasm (Gundersen & Worman, 2013).

Second, the coupling between the NE and the cytoskeleton enables the *coordination of nuclear activities with cellular processes* by anchoring the nucleus to the cytoskeleton, which helps regulate nuclear shape and size. Cytoskeletal filaments serve as a scaffold that spatially connects the nuclear envelope (NE) and the plasma membrane, enabling the propagation of mechanical stimuli from the extracellular environment to the nucleus and back. To elicit a physiological response, these mechanical stimuli are transduced by two fundamental families of receptors that trigger signalling cascades.[17] The first family, integrins, located in the plasma membrane, plays a critical role in organising multiple nuclear components (e.g. chromatin and nucleoli) and influencing the dynamics and intracellular localisation of heterochromatin (Ramdas & Shivashankar, 2015; Hieda, 2019). The second family, the LINC complex (Linker of Nucleoskeleton and Cytoskeleton),[18] is situated in the NE, spanning both the inner nuclear membrane (INM) and the outer nuclear membrane (ONM). This arrangement allows signals to travel in two directions: from the cytoplasm to the nucleoplasm ("outside-in signals") and from the nucleoplasm to the cytoplasm ("inside-out signals") (Hieda, 2019).

Outside-in signals can follow two main pathways: one where mechanical stimuli are directly transferred to the nucleus via the LINC complex, and another where transcription factors (e.g. Yes-associated proteins) shuttle between the cytoplasm and the nucleus through the nuclear pore complexes (NPCs) and are regulated by their association/dissociation with actin filaments. Inside-out signals involve the transfer of signals from the nucleoplasm to the cytoplasm across the NE. The domain proteins of the LINC complex, such as the SUN-domain, are involved in these inside-out signals, playing roles in the activation/inhibition of RhoA (a class of GTPases) and in the assembly of focal adhesions[19] (Hieda, 2019).

---

[16] MTOC is a eukaryotic structure from which microtubules emerge. The NE associates with the MTOC, determining where microtubule ends anchor (Gundersen & Worman, 2013).

[17] Mechanical transduction involves transforming mechanical stimuli (e.g. pressure changes) into chemical signalling cascades known as "mechanosensors".

[18] The proteins of the LINC complex, including SUN-domain and KASH-domain proteins, interact with lamins and chromatin in the nucleus, while nesprins associate with cytoskeletal elements in the cytoplasm. Potential regulators of the LINC complex include intraluminal calcium, ubiquitination, torsinA, the redox environment of the NE, and chromatin (Hieda, 2019).

[19] Focal adhesions are macromolecular complexes that transmit mechanical force and regulatory signals between the extracellular matrix and the cell (Hieda, 2019).

Third, the NE-cytoskeleton coupling *supports the mechanotransduction processes*, allowing the nucleus to respond to mechanical stimuli and stress, which is important for cellular development, migration, and stability. The bond between the nuclear envelope (NE) and microtubules plays a crucial role in mitosis, during which replicated chromatin must be equally divided into daughter cells by the mitotic spindle. Normally excluded from the nucleus during interphase, microtubules must gain access to chromatin for mitosis to proceed. Motor proteins, such as dynein, exert mechanical forces on the NE, deforming it and promoting its breakdown, which allows microtubules to interact directly with chromatin at the onset of mitosis. This NE breakdown eliminates the membrane barrier around chromatin, facilitating the formation of the mitotic spindle. At the conclusion of mitosis, the NE reassembles around the decondensing chromatin, while microtubules, left from spindle disassembly, remain in the vicinity of the chromatin area, aiding in the reestablishment of the NE structure (De Magistris & Antonin, 2018).

To conclude, the NE is a fundamental *hinge* between the *nucleoplasm* and the *cytoplasm* that reveals key aspects of the physiological integration between them. First, the NE enables *selective molecular traffic* into and out of the nucleus, allowing essential components of transcriptional processes to be imported while exporting transcripts. This selective permeability ensures effective communication across the cell and provides crucial regulatory control over genetic and metabolic processes, fundamental to eukaryotic physiology. Secondly, the NE's strong linkage with cytoskeletal filaments permits the nucleus to *change its position* within the cell and facilitates the reception and generation of signals that are exchanged with the nucleoplasm. This interaction coordinates nuclear functions with other eukaryotic subregions. Thirdly, *the cytoskeleton constrains and displaces the nucleus* to meet the cell's physiological needs. During mitosis, for instance, the cytoskeleton temporarily disrupts the NE to facilitate the division of chromatin into daughter cells, demonstrating its role in cellular replication and division.

6.5 The Eukaryotic Endomembrane System and the Inter-Organelle Communication

The eukaryotic endomembrane system consists of a series of interconnected organelles, each performing a fundamental role in the physiology of the eukaryotic cell: the endoplasmic reticulum (ER), the Golgi apparatus, lysosomes, and endosomes[20] (Fig. 6.3). These organelles communicate with each other through direct contact at membrane contact sites and vesicular transport over longer distances. This section examines three organisational aspects of the eukaryotic endomembrane system: the

[20] It is noteworthy that plastid and mitochondrial membranes, despite functional connections to the endomembrane system, are not part of it. All eukaryotic endomembrane system organelles are involved in secretory pathways, forming a single functional unit.

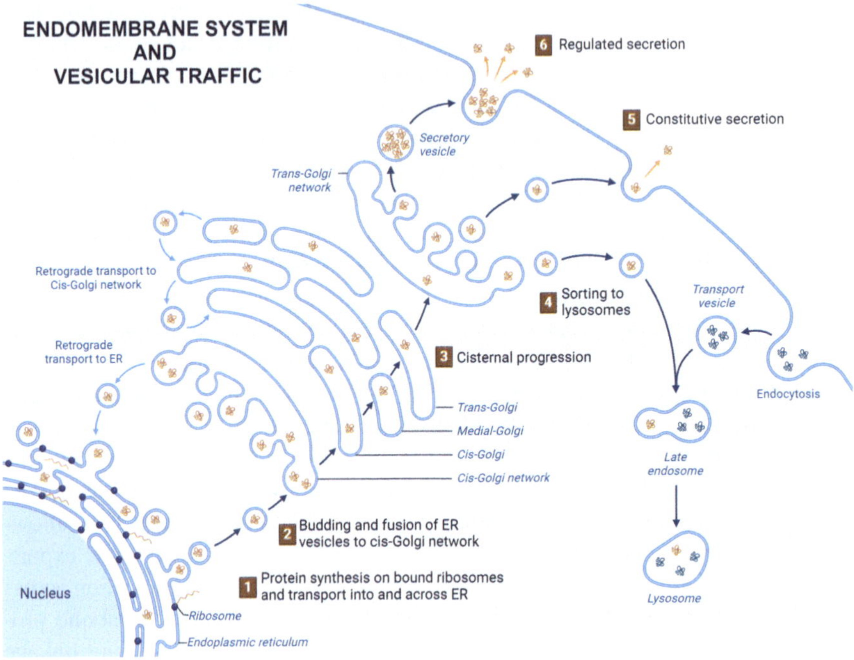

Fig. 6.3 The endomembrane system and vesicle transport. This figure illustrates the interconnected network of organelles within the eukaryotic cell that make up the endomembrane system, including the endoplasmic reticulum (ER), Golgi apparatus, lysosomes, and vesicles. It highlights the process of vesicle transport, where proteins and lipids are synthesised in the ER, modified and sorted in the Golgi apparatus, and then transported via vesicles to their final destinations, such as the plasma membrane, lysosomes, or other organelles. This system is crucial for maintaining cellular organisation, communication, and transport of materials within the cell

structure and functional contributions of the ER, Golgi apparatus, lysosomes, and endosomes; the role of contact sites and vesicular transport in inter-organelle communication; and how the eukaryotic cell functionally sustains the organelles of the endomembrane system.

The endoplasmic reticulum (ER) is the largest organelle in eukaryotic cells, consisting of two main regions: the rough ER and the smooth ER. The rough ER, whose membrane is contiguous with the outer nuclear membrane, is characterised by the presence of numerous ribosomes on its surface. These ribosomes are crucial for protein biogenesis, folding, and post-translational modifications (Braakman & Hebert, 2013). In contrast, the smooth ER lacks ribosomes due to its highly curved and smooth surface (Schwarz & Blower, 2016). Consequently, the smooth ER is not involved in protein biogenesis but is essential for lipid synthesis[21] and transport (Fagone & Jackowski, 2009), calcium storage (Clapham, 2007), carbohydrate

[21] Phospholipid synthesis occurs in a region rich in tubules and vesicles between the ER and the Golgi apparatus (Fagone & Jackowski, 2009).

metabolism (Hebert et al., 2005), and signalling between the ER and other organelles (Westrate et al., 2015).

The Golgi apparatus is located adjacent to the ER and consists of a series of membrane-enclosed disks that form a stacked cisternal structure with distinct polarity: the cis-face, which receives material from the ER, and the trans-face, which directs material towards lysosomes, secretory vesicles, and the cell membrane (Day et al., 2013). The Golgi apparatus has two primary functions. Firstly, it is involved in significant post-translational modifications, such as the addition or removal of carbohydrates from proteins synthesised in the ER. Secondly, the Golgi complex receives many secretory proteins from the ER, packages them using various signalling mechanisms,[22] and dispatches them to specific subcellular destinations (Short & Barr, 2000). The exact mechanisms of vesicle transport within the Golgi apparatus are not entirely understood, but two main models have been proposed: vesicle transport and cisternal maturation. According to the vesicle transport model, cargo molecules move through the pre-existing Golgi cisternae by fusing with the cis-face and releasing their contents into the cisternae, then forming new vesicles to reach other locations within the Golgi apparatus. The cisternal maturation model suggests that ER vesicles coalesce to form the cis-face cisterna, which then matures by progressively acquiring and releasing enzymes and other components as it moves through the Golgi stack (Jackson, 2009).

Endosomes and lysosomes form a unique system of spherical organelles characterised by an acidic intracellular environment that breaks down fundamental macromolecules into their constituent components. Endosomes are categorised into three types: early, recycling, and late endosomes. Early endosomes function as major sorting stations, receiving vesicles filled with proteins from the plasma membrane. Recycling endosomes return coated vesicles to the cell surface, while late endosomes receive these vesicles and fuse with lysosomes to initiate the proteolytic process (Hu et al., 2015). Lysosomes contain numerous enzymes (hydrolases) that break down proteins, lipids, and carbohydrates into their building blocks—amino acids, fatty acids, and monosaccharides, respectively. Additionally, lysosomes possess various ion channels (e.g. Ca^{2+}, Cl^-, H^+) and transporters (e.g. amino acid, lipid, sugar, and heavy metal exporters) that enable them to receive both extracellular cargos via endocytic pathways and intracellular components through autophagy.

The organelles of the endomembrane system perform diverse functions and produce various molecules that must be exchanged among different microcompartments in a highly coordinated and regulated manner. One strategy for inter-organelle communication involves *membrane contact sites*,[23] which are domains where the

---

[22] Usually, these signals are induced by GTPases.

[23] Membrane contact sites feature four distinct types of proteins: structural, functional, regulator, and sorter proteins. Structural proteins (tethers) hold two distinct organelles together at a defined distance (spacers). Functional proteins (e.g. ion channels and pumps, lipid transfer proteins, metabolite channels, and transporters) facilitate molecule passage across the membrane contact site. Regulator proteins modulate the behaviour of other membrane contact site proteins through mechanisms like phosphorylation or redox state changes. Cellular metabolism controls these regu-

membranes of the ER are adjacent to those of other organelles, such as mitochondria, lysosomes, endosomes, the Golgi apparatus, lipid droplets, and the plasma membrane. These contact sites facilitate the exchange of proteins, lipids, and calcium between the organelles (Helle et al., 2013). A crucial aspect of membrane contact sites is their role in transmitting Ca^{2+} to mitochondria and the plasma membrane,[24] thereby acting as signalling platforms for regulating organelle biogenesis, dynamics, inheritance, positioning, fission, and autophagy (Wu et al., 2018; Scorrano et al., 2019).

Another fundamental strategy employed by organelles to communicate among each other is the use of *vesicles* for transporting lipids and proteins (Fig. 6.3). Vesicles are produced by a donor organelle and fuse with an acceptor in another part of the cell, thus passing through membrane barriers without altering the functional organisation of the organelles[25] (Gomez-Navarro & Miller, 2016). Vesicles can move in two different directions: they either flow from the ER to the Golgi apparatus to the plasma membrane (the exocytic pathway) or from the plasma membrane to endosomes/lysosomes, then to the Golgi complex, and back to the ER (the endocytic pathway). Vesicles can be classified into three main groups: COPII,[26] COPI, and clathrins. COPII vesicles transport lipids and proteins from the ER to the Golgi apparatus (anterograde transport), COPI vesicles transport materials from the Golgi to the ER (retrograde transport) and between Golgi cisternae, and clathrin-coated vesicles form at the plasma membrane and in the Golgi apparatus to fuse with endosomes or lysosomes. Vesicles are filled with proteins and lipids through the action of various molecules (e.g. adaptors, receptors, and accessory factors) that, after receiving signals from cargo proteins or lipids, recruit these cargoes into vesicles (Geva & Schuldiner, 2014; Gomez-Navarro & Miller, 2016).

Vesicular transport is tightly regulated to ensure highly *coordinated* and *integrated movement* of molecules among organelles. Small GTPases[27] play a critical role in this regulation by overseeing the maturation of Golgi cisternae and endosomes, coordinating vesicular transport steps, and integrating these processes into cohesive pathways (Segev, 2011). Although the precise mechanisms remain partially elucidated, small GTPases are believed to facilitate the maturation of Golgi

latory proteins, establishing a fundamental functional dependence of membrane contact sites on overall cell physiology. For example, the contact site between mitochondria and vacuoles (vCLAMP) is regulated by the phosphorylation of Vps39 via cellular kinase cascades (Hönscher et al., 2014). Sorter proteins recruit proteins into the contact site (Scorrano et al., 2019).

[24] In eukaryotic multicellular organism cells, ER membrane contact sites also transmit apoptotic signals to mitochondria (Wu et al., 2018).

[25] Although the regulation of vesicle biogenesis is not fully understood, cargos are believed to allosterically regulate the synthesis of coated vesicles (Gomez-Navarro & Miller, 2016).

[26] COP stands for Coat Protein Complex, which comprises four different protein subunits involved in the budding process.

[27] GTPases are a family of hydrolase enzymes activated by binding to guanosine triphosphate (GTP). Small GTPases act as molecular switches and signal transducers in many cellular signalling processes. In humans, the Rab family of GTPases plays a crucial role in vesicle transport regulation.

cisternae and endosomes by recruiting specific protein effectors[28] that alter the protein composition of these compartments. These GTPases interact with protein effectors to ensure that vesicles emerging from one compartment (e.g. the ER) are accurately targeted to and fuse with the appropriate organelle (e.g. the Golgi apparatus) (Segev, 2011). Additionally, small GTPases are instrumental in integrating individual transport steps into comprehensive exocytic and endocytic pathways.[29] They are activated by proteins located on the organelle membranes, such as the TRAPP protein complexes[30] and GEF[31] proteins on the Golgi membrane, which bind to small GTPases and trigger the regulation of vesicular transport (Lipatova & Segev, 2019).

The overall functioning of the organelles in the endomembrane system is critically dependent on *global cellular conditions*, such as growth, survival, and homeostasis (Farhan & Rabouille, 2011). Extracellular signals[32] binding to cell membrane receptors[33] initiate cytosolic signalling pathways.[34] These pathways, in turn, interact with transmembrane receptors on organelles, triggering signalling cascades that remodel organelle architecture[35] and regulate the functions of organelles within the secretory pathway.

For example, protein export from the ER secretory pathway is modulated by signalling cascades, such as MAP kinases, which are activated in response to extracellular signals. These cascades lead to the phosphorylation of ER chaperones (e.g. calnexin) and COPII vesicle components (e.g. Sec16), thereby regulating cargo flux through the ER. Similarly, the organisation of Golgi cisternae and Golgi protein trafficking is influenced by signalling pathways, such as those involving p38 MAP kinases and PKA.[36] These pathways, activated by extracellular signals, phosphorylate key Golgi proteins, remodelling Golgi structure and facilitating post-Golgi trafficking (Farhan & Rabouille, 2011). Endosomes and lysosomes also respond to cellular signals, including small GTPases that sense nutrient availability, thereby regulating

[28] For example, Rab1 interacts with two protein effectors, p115 and GM130 (Segev, 2011).

[29] For example, the yeast exocytic pathway's two main steps, from ER to Golgi and Golgi to the plasma membrane, are integrated into one pathway through the interaction between Rab Ypt and the modular GEF complex TRAPP (Segev, 2011, p. 37).

[30] TRAPP, or *Transport Protein Particle*, is a multi-subunit protein that binds to ER-derived vesicles, bringing them closer to the membrane acceptor. Two TRAPP families, TRAPP I and TRAPP II, have been identified (Sacher et al., 2008).

[31] GEF (Guanine Nucleotide Exchange Factor) is a protein that activates GTPases by promoting the binding of GTPases to GTP.

[32] Examples include growth factors or cytokines involved in cell growth, survival, and homeostasis (Farhan & Rabouille, 2011).

[33] Examples include cytokine receptors and receptor tyrosine kinases.

[34] Key molecules in signalling cascades include kinases (e.g. the mitogen-activated protein (MAP) kinase pathway), phosphatases, GTPases, nucleotides, and lipid mediators (Farhan & Rabouille, 2011).

[35] Organelle architecture remodelling (membrane biogenesis) refers to regulating organelle quantity and composition according to cellular needs (Nunnari & Walter, 1996).

[36] PKA stands for Protein Kinase A.

lysosomal catabolic processes (Settembre et al., 2013). Consequently, the endo-lysosomal system plays a crucial role in maintaining cellular homeostasis and regulating essential cellular functions, including growth and apoptosis (Xu & Ren, 2015).

It is important to recognise that the organelles of the endomembrane system not only *receive* but also *generate* and *modulate cellular signals*, leading to systemic physiological responses within the eukaryotic cell. For instance, MAP kinase cascades interact with proteins located on the membranes of the ER (e.g. Ras) and Golgi apparatus (e.g. Ras, Raf, dynamins). These interactions trigger signals that are transmitted to the plasma membrane, thereby influencing processes such as cell growth, proliferation, and secretion.

In summary, three key organisational dimensions define the functional integration of the endomembrane system organelles:

1. *Inter-Organelle Communication:* The functions of the ER, Golgi apparatus, endosomes, and lysosomes are interconnected through inter-organelle communication. This involves signalling cascades, such as those mediated by GTPases and MAP kinases, which are activated by proteins on the organelle membranes.
2. *Dependency on Cellular Physiology:* The mechanisms of the secretory pathway organelles are closely tied to the overall cell physiology. They are regulated by environmental and cytosolic signals that bind to organelle membrane receptors. This signalling informs the organelles about extracellular and intracellular conditions, enabling them to respond appropriately.
3. *Impact on Eukaryotic Cell Physiology:* The endomembrane system is crucial for eukaryotic cell physiology in two fundamental ways. First, the organelles' mechanisms directly influence cellular functions. Second, the signals transmitted from these organelles to the plasma membrane modulate cellular behaviours and responses.

6.6 Functional Integration in Prokaryotic and Eukaryotic Membrane-Bound Compartments

This section compares prokaryotic and eukaryotic membrane-bound compartments to highlight their distinct contributions to the functional organisation and integration of unicellular organisms. The case studies demonstrate how variations in the organisation of intracellular spaces, through different types of internal membranes, uniquely affect the overall physiology of the organism. Moreover, they illustrate that the internal spatial organisation within an organism dictates how the functions of its component parts interact and integrate.

Interestingly, this organisational feature of unicellular organisms is also evident in multicellular ones. As Bich et al. (2019) argue, the internal space of multicellular organisms is functionally organised. Specifically, the extracellular matrix (ECM) structures the intercellular space and acts as a spatial constraint on cellular functions. This organisation is crucial for fundamental properties such as intercellular communication, cell behaviour, and cell fate.

Prokaryotic and eukaryotic endomembranes share the function of dividing the intracellular space into distinct micro-regions, each serving as a unique biochemical micro-environment. These compartments are physically separated from the cytoplasm but remain physiologically connected to it. This separation and connection have profound physiological consequences. For instance, the isolation of specific biochemical reactions enables processes that would otherwise be impossible, such as those involving toxic metabolic intermediates in prokaryotic microcompartments. Additionally, this compartmentalisation allows for more complex regulatory mechanisms, exemplified by the distinct regulation of transcription and translation in different cellular compartments. Furthermore, it facilitates the emergence of new and more complex biochemical pathways, such as the vesicles and the secretory pathway in eukaryotic cells.

A key difference between prokaryotic and eukaryotic microcompartments lies in the range of biological functions they perform. Prokaryotic lipid- or protein-bounded microcompartments primarily *isolate certain metabolic intermediates*, which would otherwise be toxic or too volatile in the cytoplasm, to carry out various metabolic (notably catabolic) reactions. For instance, carboxysomes facilitate crucial reactions of the Calvin cycle, magnetosomes enable bacteria to align with the Earth's magnetic field, and the anammoxosomes in planctomycetes perform anaerobic ammonium oxidation. In these cases, prokaryotic microcompartments optimise metabolic reactions by isolating specific intermediates and enzymes. Essentially, prokaryotic microcompartments extend the cytoplasm-based metabolism without being directly involved in the synthesis of constitutive cellular components (such as RNA molecules, proteins, and lipids) or in their genetic regulation.

In contrast, eukaryotic organelles like the nucleus and the endomembrane system are intricately involved in *metabolic pathways* (e.g. lipid and cholesterol synthesis in the smooth endoplasmic reticulum) as well as crucial processes such as *transcription* (in the nucleus), *translation* (by ribosomes on the rough endoplasmic reticulum), *post-transcriptional modification* (in the nucleus), and *post-translational modification* (in rough endoplasmic reticulum and in the Golgi complex). This distinction highlights a significant organisational difference between prokaryotic and eukaryotic organelles: eukaryotic organelles play a vital role in the *synthesis and regulation of the cell's constitutive components* (proteins and lipids). The metabolic pathways in the cytoplasm depend on the presence of specific proteins (e.g. enzymes, signalling molecules, receptors, transporters, and pumps), whose synthesis and genetic regulation are governed by the nucleus and endomembrane system. This level of control over cellular processes is absent in prokaryotic microcompartments, illustrating the complex regulatory capabilities of eukaryotic cells.

The different functions and types of control exerted by prokaryotic and eukaryotic organelles lead to distinct methods of physiological integration within the cellular network. The functional integration between prokaryotic microcompartments and the prokaryotic cell exhibits at least three organisational features. First, the biogenesis of microcompartment membranes is dependent on the *genetic machinery* located in the cytoplasm, creating a direct causal relationship between the organelle's structure and the cell's genetic machinery (Murat et al., 2010). Second, the metabolic functions of microcompartments rely on the correct *transport of metabolites* from the cytoplasm to the microcompartment's lumen. This transport is facilitated by the intrinsic physicochemical properties of intracellular membranes and their transmembrane proteins (Bobik et al., 2015). Third, the metabolic reactions within prokaryotic microcompartments produce essential *metabolic products* for the overall cellular metabolism, contributing to the self-maintenance and constitutive functions of prokaryotic cells (Murat et al., 2010).

Since prokaryotic microcompartments do not engage in transcription or translation, they do *not* require a complex protein machinery for genetic regulation. Moreover, the absence of multiple microcompartments within the same cell eliminates the need for a sophisticated intracellular communication system in prokaryotic cells. This simplicity contrasts with eukaryotic cells, where organelles perform specialised roles and require intricate communication and regulation mechanisms to coordinate their functions.

Compared to prokaryotic microcompartments, the functional integration between the nucleus and the cytoplasm in eukaryotic cells is significantly more complex, involving intricate *signalling* and *regulatory pathways*. A major challenge is the coordination of genetic transcription with numerous other cellular processes, necessitating the interplay between several entities.

First, the cytoskeleton plays a critical role by not only positioning the nucleus but also transmitting mechanical stimuli between the nucleoplasm and the cytoplasm (Tapley & Starr, 2013). Second, transmembrane receptors located on the NE and the cellular membrane relay cytoplasmic signals that can either trigger or inhibit genetic transcription based on cellular needs (Hieda, 2019). Third, the nuclear protein import/export machinery enables the export of RNA molecules to ribosomes and the import of enzymes and regulatory proteins into the nucleus, facilitating the transcription and translation processes (Görlich & Kutay, 1999).

These entities work in a coordinated manner to ensure functional integration between the nucleus and the rest of the eukaryotic cell. Firstly, cytoplasmic signals and the nuclear import/export machinery must exhibit a high level of coordination to regulate the flow of molecules necessary for transcription and translation according to physiological needs (Hieda, 2019). This regulation ensures that the genetic information is accurately transcribed and translated into proteins as required by the cell's condition.

Additionally, the positioning and movement of the nucleus by the cytoskeleton likely enhance the transmission of signals between the cytoplasm and the nucleoplasm. This improved signalling facilitates the efficient coordination of transcription and translation, ensuring that the cellular responses are timely and appropriate to the stimuli received (Tapley & Starr, 2013).

In summary, the functional integration between the nucleus and cytoplasm in eukaryotic cells involves a sophisticated network of *signalling pathways*, *mechanical interactions*, and *molecular transport mechanisms*, all of which are essential for maintaining cellular homeostasis and responding to environmental changes.

Membrane contact sites and vesicles play a fundamental role in the functional integration between the endomembrane system and the cytoplasm by facilitating the exchange of proteins and lipids among the ER, the Golgi complex, and the cytoplasm. Membrane contact sites allow for direct communication and substance transfer between adjacent organelles, while vesicles transport materials over longer distances within the cell.

The release of substances at membrane contact sites (Scorrano et al., 2019) and via vesicles (Gomez-Navarro & Miller, 2016) is tightly regulated by a variety of signals, including GTPases, which coordinate the metabolic and regulatory processes within the organelles with those in the cytoplasm. These signals ensure that metabolic intermediates and regulatory molecules are correctly distributed, maintaining cellular homeostasis.

Additionally, the structure and functions of the organelles within the endomembrane system are modulated by signalling pathways, such as MAP kinases, which are activated by extracellular signals (Farhan & Rabouille, 2011). These pathways enable the cell to respond dynamically to external stimuli, adjusting organelle functions to meet cellular demands. This coordination is essential for the cell's ability to adapt to changing environmental conditions and to maintain its overall functionality.

To conclude, I will provide a more abstract and general account of the functional integration between microcompartments and the cell by employing a basic definition of cellular functional integration as offered in Chap. 3. There, I proposed a fundamental definition of cellular physiological integration in terms of an organisational closure between S (signalling pathways), R (regulatory processes), and C (the constitutive processes of the cell). We may now ask whether this closure is maintained in the presence of organelles and, if so, how.

In the case of prokaryotic microcompartments, since they primarily isolate catabolic reactions without synthesising constitutive components (e.g. proteins or lipids) of the cell, they only marginally contribute to C. Additionally, they do not perform genetic regulation and their membranes do not appear to be involved in a complex signalling network with the receptors of the cellular membrane. Consequently, the functional integration of prokaryotic microcompartments in the cell does not entail a profound transformation of the basic organisational closure of the cell between C, R, and S.

In prokaryotic cells, most of the closure between C, R, and S occurs at the *interface* between the *plasma membrane* and the *cytoplasm*. The plasma membrane plays a central role in sensing environmental signals and initiating regulatory pathways, which in turn affect cytoplasmic processes. The metabolic activities within the prokaryotic microcompartments, while important, represent only a minor part of the overall cellular metabolism and are largely independent of the complex regulatory and signalling networks that characterise eukaryotic cells.

Therefore, while prokaryotic microcompartments contribute to the cell's metabolic efficiency by isolating specific reactions and intermediates, they do not significantly alter the fundamental organisational structure of the cell. The major processes that ensure cellular integrity and functionality—signalling, regulation, and the synthesis of cellular components—remain predominantly associated with the plasma membrane and the cytoplasm, rather than with the microcompartments. This delineation underscores a key difference between prokaryotic and eukaryotic cells in terms of the integration and complexity of their internal compartments.

In eukaryotes, the closure between C, R, and S becomes highly complex for two main reasons.

First, the patterns of interaction between C, R, and S are no longer localised only (or mostly) in the cytoplasm, but rather are shared between the cytoplasm, the nuclear envelope, the nucleoplasm, and the membranes and organelles of the endomembrane system. Eukaryotic organelles are involved in *constitutive processes* (e.g. biosynthesis of RNA molecules, proteins, and lipids), their *regulation* (e.g. transcriptional, post-transcriptional, translational, and post-translational regulation), and a variety of *signalling pathways* (e.g. signalling pathways for nuclear import/export or the secretory pathway) that regulate biochemical processes occurring within the organelle lumen.

Secondly, to ensure the physiological integration of C, R, and S among different cellular sub-regions, *inter-organelle communication* is required. This complex form of functional coordination is achieved through a variety of signalling pathways (e.g. the Ras signalling pathway, the cAMP-PKA pathway, heterotrimeric G proteins), the receptors for which are located within the plasma membrane and the membranes of organelles. Inter-organelle communication is fundamental to eukaryotic physiology: it coordinates membrane fluxes along each of the trafficking segments (e.g. endoplasmic reticulum-Golgi complex, Golgi complex-plasma membrane, nuclear envelope-plasma membrane, nuclear envelope-mitochondria) to avoid gross imbalances in the system. Additionally, trafficking fluxes respond to extracellular signals, thus adapting the interior of the cell to extracellular requests (Sallese et al., 2006).

This intricate network of interactions ensures that the activities of the nucleus, endoplasmic reticulum, Golgi apparatus, and other organelles are *synchronised* with cellular needs and external conditions. For instance, the endoplasmic reticulum is involved in the synthesis of proteins and lipids, which are then processed and sorted in the Golgi apparatus before being transported to their final destinations. The nucleus, which houses the cell's genetic material, is constantly communicating with the cytoplasm to regulate gene expression in response to cellular signals.

In summary, eukaryotic cells exhibit a sophisticated level of functional integration and coordination among their organelles, far surpassing that of prokaryotic cells. This complexity enables eukaryotic cells to perform a wide range of specialised functions and respond dynamically to their environment, thereby supporting the advanced physiological processes characteristic of multicellular organisms.

6.7 Concluding Remarks

In this chapter, I have explored the structural and functional organisation of prokaryotic and eukaryotic membrane-bound compartments, and examined the key physiological mechanisms underpinning their functional integration within the cell. The case studies reveal a fundamental difference between prokaryotic and eukaryotic organelles: prokaryotic microcompartments primarily isolate and metabolically transform substances that would otherwise be toxic or volatile in the cytoplasm. In contrast, eukaryotic organelles are crucially involved in the synthesis and regulation of the cell's constitutive components, such as RNA molecules, proteins, and lipids. This distinction highlights the broader complexity and functional diversity in eukaryotic cells compared to their prokaryotic counterparts, reflecting the advanced organisational structures and regulatory networks required for maintaining cellular function and adaptation in more complex organisms.

This functional diversity leads to distinct mechanisms for the physiological integration of organelles within the cell. In prokaryotic microcompartments, the synthesis and regulation of their components are managed by the cytoplasmic genetic machinery. Their metabolic functions are primarily dependent on how intracellular membranes regulate the flow of metabolic intermediates to and from the cytoplasm.

In eukaryotic cells, integration between organelles and the cytoplasm involves a complex network of signalling pathways. Receptors for these pathways are present in both the plasma membrane and the membranes of various organelles. Additionally, the cytoskeleton plays a crucial role in this integration. It not only transmits mechanical stimuli from the cytoplasm to the organelles but also facilitates the movement of organelles in response to physiological demands, thereby ensuring a dynamic and responsive cellular organisation.

I have argued that the closure between C, R, and S is maintained in both prokaryotic and eukaryotic cells, though it is organised differently in prokaryotic and eukaryotic microcompartments. In prokaryotic cells, this closure predominantly occurs within the cytoplasm. Microcompartments in prokaryotes function as extensions of the cytoplasmic processes, without fundamentally altering the organisation of C, R, and S.

In contrast, eukaryotic cells distribute C, R, and S across various organelles. This distribution means that the constitutive dimension of eukaryotic autonomy arises from the combined physiological contributions of both organelles and the cytoplasm, rather than being confined to the cytoplasm alone. This significant organisational shift is underpinned by intricate intracellular communication mechanisms that enable organelles to function in concert with each other and with the cytoplasm, thereby achieving a more complex and coordinated cellular system.

The difference in the organisation of C, R, and S in prokaryotic and eukaryotic membrane-bound compartments raises the crucial question of why these organisms exhibit such distinct organisational strategies and what evolutionary possibilities each configuration affords.

Prokaryotic membrane-bound compartments are believed to have evolved independently multiple times, as evidenced by the distinct proteins that govern the formation of each organelle system. These compartments are thought to originate from inward folding of the plasma membrane, through processes such as tubulation, invagination, or vesiculation, indicating an autogenous origin (Murat et al., 2010). Two notable points arise from this: first, because membrane-bound compartments are not universally present in all bacteria and archaea but are specific to certain lineages, their evolution likely reflects particular *metabolic demands*. These demands may have driven the evolution of specialised functions by either repurposing existing proteins or by limited protein innovation. Second, the relatively small number of microcompartments in prokaryotic cells, each with specialised roles, suggests *systemic constraints*—such as cell size and bioenergetic limitations—that restrict the development of a diverse array of internal membranes. This constraint results in a simpler, more constrained system compared to the eukaryotic cellular complexity.

While the endosymbiotic origin of mitochondria and chloroplasts is widely accepted (Archibald, 2009; Martin et al., 2015; Raval et al., 2022), the evolutionary origins of the nuclear envelope and the endomembrane system remain a subject of ongoing debate. As discussed in Chap. 4, three primary theories have been proposed to explain their origins: the endosymbiotic theory, the autogenous theory, and the inside-out model. According to the endosymbiotic theory (Gupta & Bryan Golding, 1996), the nuclear envelope and other endomembranes evolved from endosymbiotic events similar to those that gave rise to mitochondria and chloroplasts. In contrast, the autogenous theory (Cavalier-Smith, 1975) posits that these structures originated from a series of internal modifications—such as invagination, tubulation, and vesiculation—of the plasma membrane of a prokaryotic ancestor. The inside-out model (Baum & Baum, 2014) suggests that the outer nuclear membrane evolved from protrusions of the plasma membrane, with the endoplasmic reticulum forming from the spaces between these protrusions.

In light of the above discussion, several observations and hypotheses can be proposed regarding the evolution and functional significance of intracellular membranes in eukaryotes.

Firstly, the ubiquitous presence of intracellular membranes in eukaryotic cells aligns with their critical roles in essential functions such as *genetic regulation* and *metabolic processes*. These functions are fundamental to eukaryotic life, indicating that intracellular membranes are integral to maintaining cellular complexity and efficiency.

Secondly, the development of intracellular membranes is closely linked to the emergence of sophisticated signalling pathways that facilitate *intracellular communication*. By compartmentalising functions that were previously managed in the cytoplasm, eukaryotic cells have enhanced their biochemical processes. This compartmentalisation allows for more precise regulation of key processes, such as transcription and translation, thereby expanding the physiological capabilities of the cell.

Finally, the proliferation of organelles in eukaryotic cells likely co-evolved with the *increase in cell size*. As eukaryotic cells became larger, the division of internal space into specialised compartments may have optimised genetic and metabolic functions to meet increased bioenergetic demands (Lane, 2014). Additionally, the physiological needs of larger cells could have driven the functional diversification of organelles, contributing to the wide array of functional innovations observed in eukaryotic cells.

References

Almeida, N. M. d., Neumann, S., Mesman, R. J., Ferousi, C., Keltjens, J. T., Jetten, M. S. M., Kartal, B., & van Niftrik, L. (2015). Immunogold localization of key metabolic enzymes in the Anammoxosome and on the Tubule-like structures of Kuenenia Stuttgartiensis. *Journal of Bacteriology, 197*(14), 2432–2441. https://doi.org/10.1128/JB.00186-15

Archibald, J. M. (2009). The puzzle of plastid evolution. *Current Biology, 19*(2), R81–R88. https://doi.org/10.1016/j.cub.2008.11.067

Baum, D. A., & Baum, B. (2014). An inside-out origin for the eukaryotic cell. *BMC Biology, 12*, 76. https://doi.org/10.1186/s12915-014-0076-2

Bernardino de la Serna, Jorge, G. J., Schütz, C. E., & Cebecauer, M. (2016). There is no simple model of the plasma membrane organization. *Frontiers in Cell and Developmental Biology, 4*, 106. https://doi.org/10.3389/fcell.2016.00106

Bich, L., Pradeu, T., & Moreau, J.-F. (2019). Understanding multicellularity: The functional organization of the intercellular space. *Frontiers in Physiology, 10*, 1170. https://doi.org/10.3389/fphys.2019.01170

Bobik, T. A., Lehman, B. P., & Yeates, T. O. (2015). Bacterial microcompartments: Widespread prokaryotic organelles for isolation and optimization of metabolic pathways. *Molecular Microbiology, 98*(2), 193–207. https://doi.org/10.1111/mmi.13117

Braakman, I., & Hebert, D. N. (2013). Protein folding in the endoplasmic reticulum. *Cold Spring Harbor Perspectives in Biology, 5*(5), a013201. https://doi.org/10.1101/cshperspect.a013201

Cavalier-Smith, T. (1975). The origin of nuclei and of eukaryotic cells. *Nature, 256*(5517), 463–468. https://doi.org/10.1038/256463a0

Chowdhury, C., Chun, S., Pang, A., Sawaya, M. R., Sinha, S., Yeates, T. O., & Bobik, T. A. (2015). Selective molecular transport through the protein Shell of a bacterial microcompartment organelle. *Proceedings of the National Academy of Sciences of the United States of America, 112*(10), 2990–2995. https://doi.org/10.1073/pnas.1423672112

Chowdhury, C., Sinha, S., Chun, S., Yeates, T. O., & Bobik, T. A. (2014). Diverse bacterial microcompartment organelles. *Microbiology and Molecular Biology Review, 78*(3), 438–468. https://doi.org/10.1128/MMBR.00009-14

Clapham, D. E. (2007). Calcium Signaling. *Cell, 131*(6), 1047–1058. https://doi.org/10.1016/j.cell.2007.11.028

Cornejo, E., Abreu, N., & Komeili, A. (2014). Compartmentalization and organelle formation in bacteria. *Current Opinion in Cell Biology, 26*, 132–138. https://doi.org/10.1016/j.ceb.2013.12.007

Cybulski, L. E., Martín, M., Mansilla, M. C., Fernández, A., & de Mendoza, D. (2010). Membrane thickness Cue for cold sensing in a bacterium. *Current Biology, 20*(17), 1539–1544. https://doi.org/10.1016/j.cub.2010.06.074

D'Angelo, M. A., & Hetzer, M. W. (2006). The role of the nuclear envelope in cellular organization. *Cellular and Molecular Life Sciences, 63*(3), 316–332. https://doi.org/10.1007/s00018-005-5361-3

Day, K. J., Andrew Stahelin, L., & Glick, B. S. (2013). A three-stage model of Golgi structure and function. *Histochemistry and Cell Biology, 140*(3), 239–249. https://doi.org/10.1007/s00418-013-1128-3

De Magistris, P., & Antonin, W. (2018). The dynamic nature of the nuclear envelope. *Current Biology, 28*(8), R487–R497. https://doi.org/10.1016/j.cub.2018.01.073

Devos, D. P., Gräf, R., & Field, M. C. (2014). Evolution of the nucleus. *Current Opinion in Cell Biology, 28*(100), 8–15. https://doi.org/10.1016/j.ceb.2014.01.004

Diekmann, Y., & Pereira-Leal, J. B. (2013). Evolution of intracellular compartmentalization. *The Biochemical Journal, 449*(2), 319–331. https://doi.org/10.1042/BJ20120957

Edidin, M. (2003). Lipids on the frontier: A century of cell-membrane bilayers. *Nature Reviews. Molecular Cell Biology, 4*(5), 414–418. https://doi.org/10.1038/nrm1102

Ernst, R., Ballweg, S., & Levental, I. (2018). Cellular mechanisms of physicochemical membrane homeostasis. *Current Opinion in Cell Biology, 53*, 44–51. https://doi.org/10.1016/j.ceb.2018.04.013

Fagone, P., & Jackowski, S. (2009). Membrane phospholipid synthesis and endoplasmic reticulum function. *Journal of Lipid Research, 50*(Suppl), S311–S316. https://doi.org/10.1194/jlr.R800049-JLR200

Farhan, H., & Rabouille, C. (2011). Signalling to and from the secretory pathway. *Journal of Cell Science, 124*(Pt 2), 171–180. https://doi.org/10.1242/jcs.076455

Gabaldón, T., & Pittis, A. A. (2015). Origin and evolution of metabolic sub-cellular compartmentalization in eukaryotes. *Biochimie, 119*, 262–268. https://doi.org/10.1016/j.biochi.2015.03.021

Gavrilov, A. A., & Razin, S. V. (2015). Compartmentalization of the cell nucleus and spatial organization of the genome. *Molecular Biology, 49*(1), 21–39. https://doi.org/10.1134/S0026893315010033

Geva, Y., & Schuldiner, M. (2014). The Back and forth of cargo exit from the endoplasmic reticulum. *Current Biology, 24*(3), R130–R136. https://doi.org/10.1016/j.cub.2013.12.008

Gomez-Navarro, N., & Miller, E. (2016). Protein sorting at the ER-Golgi Interface. *The Journal of Cell Biology, 215*(6), 769–778. https://doi.org/10.1083/jcb.201610031

Goñi, F. M. (2014). The basic structure and dynamics of cell membranes: An update of the singer-Nicolson model. *Biochimica et Biophysica Acta, 1838*(6), 1467–1476. https://doi.org/10.1016/j.bbamem.2014.01.006

Görlich, D., & Kutay, U. (1999). Transport between the cell nucleus and the cytoplasm. *Annual Review of Cell and Developmental Biology, 15*, 607–660. https://doi.org/10.1146/annurev.cellbio.15.1.607

Grant, C. R., Wan, J., & Komeili, A. (2018). Organelle formation in bacteria and archaea. *Annual Review of Cell and Developmental Biology, 34*, 217–238. https://doi.org/10.1146/annurev-cellbio-100616-060908

Gundersen, G. G., & Worman, H. J. (2013). Nuclear positioning. *Cell, 152*(6), 1376–1389. https://doi.org/10.1016/j.cell.2013.02.031

Gupta, R. S., & Bryan Golding, G. (1996). The origin of the eukaryotic cell. *Trends in Biochemical Sciences, 21*(5), 166–171.

Hebert, D. N., Garman, S. C., & Molinari, M. (2005). The glycan code of the endoplasmic reticulum: Asparagine-linked carbohydrates as protein maturation and quality-control tags. *Trends in Cell Biology, 15*(7), 364–370. https://doi.org/10.1016/j.tcb.2005.05.007

Helle, S. C. J., Kanfer, G., Kolar, K., Lang, A., Michel, A. H., & Kornmann, B. (2013). Organization and function of membrane contact sites. *Biochimica et Biophysica Acta, 1833*(11), 2526–2541. https://doi.org/10.1016/j.bbamcr.2013.01.028

Hieda, M. (2019). Signal transduction across the nuclear envelope: Role of the LINC complex in bidirectional signaling. *Cells, 8*(2), 124. https://doi.org/10.3390/cells8020124

Hönscher, C., Mari, M., Auffarth, K., Bohnert, M., Griffith, J., Geerts, W., van der Laan, M., Cabrera, M., Reggiori, F., & Ungermann, C. (2014). Cellular metabolism regulates contact sites between vacuoles and mitochondria. *Developmental Cell, 30*(1), 86–94. https://doi.org/10.1016/j.devcel.2014.06.006

Hu, Y.-B., Dammer, E. B., Ren, R.-J., & Wang, G. (2015). The endosomal-lysosomal system: From acidification and cargo sorting to neurodegeneration. *Translational Neurodegeneration, 4*, 18. https://doi.org/10.1186/s40035-015-0041-1

Jackson, C. L. (2009). Mechanisms of transport through the Golgi complex. *Journal of Cell Science, 122*(Pt 4), 443–452. https://doi.org/10.1242/jcs.032581

Kerfeld, C. A., Aussignargues, C., Zarzycki, J., Cai, F., & Sutter, M. (2018). Bacterial microcompartments. *Nature Reviews Microbiology, 16*(5), 277–290. https://doi.org/10.1038/nrmicro.2018.10

Kirst, H., & Kerfeld, C. A. (2019). Bacterial microcompartments: Catalysis-enhancing metabolic modules for next generation metabolic and biomedical engineering. *BMC Biology, 17*(1), 79. https://doi.org/10.1186/s12915-019-0691-z

Köhler, A., & Hurt, E. (2007). Exporting RNA from the nucleus to the cytoplasm. *Nature Reviews Molecular Cell Biology, 8*(10), 761–773. https://doi.org/10.1038/nrm2255

Lamond, A. I., & Sleeman, J. E. (2003). Nuclear substructure and dynamics. *Current Biology, 13*(21), R825–R828. https://doi.org/10.1016/j.cub.2003.10.012

Lane, N. (2014). Bioenergetic constraints on the evolution of complex life. *Cold Spring Harbor Perspectives in Biology, 6*(5), a015982. https://doi.org/10.1101/cshperspect.a015982

Lindsay, M. R., Webb, R. I., Strous, M., Jetten, M. S., Butler, M. K., Forde, R. J., & Fuerst, J. A. (2001). Cell compartmentalisation in planctomycetes: Novel types of structural organisation for the bacterial cell. *Archives of Microbiology, 175*(6), 413–429. https://doi.org/10.1007/s002030100280

Lipatova, Z., & Segev, N. (2019). Ypt/Rab GTPases and their TRAPP GEFs at the Golgi. *FEBS Letters, 593*(17), 2488–2500. https://doi.org/10.1002/1873-3468.13574

Marshall, W. F. (2012). Organelle size control systems: From cell geometry to organelle-directed medicine. *BioEssays: News and Reviews in Molecular, Cellular and Developmental Biology, 34*(9), 721–724. https://doi.org/10.1002/bies.201200043

Martin, W. F., Garg, S., & Zimorski, V. (2015). Endosymbiotic theories for eukaryote origin. *Philosophical Transactions of the Royal Society of London. Series B, Biological Sciences, 370*(1678), 20140330. https://doi.org/10.1098/rstb.2014.0330

Maturana, H. R., & Varela, F. J. (1973). *De máquinas y seres vivos: una teoría sobre la organización biológia*. Editorial Universitaria.

McInerney, J. O., Martin, W. F., Koonin, E. V., Allen, J. F., Galperin, M. Y., Lane, N., Archibald, J. M., & Martin Embley, T. (2011). Planctomycetes and eukaryotes: A case of analogy not homology. *BioEssays: News and Reviews in Molecular, Cellular and Developmental Biology, 33*(11), 810–817. https://doi.org/10.1002/bies.201100045

Militello, G., & Moreno, Á. (2018). Structural and organisational conditions for being a machine. *Biology & Philosophy, 33*(5), 35. https://doi.org/10.1007/s10539-018-9645-z

Moreno, A., & Mossio, M. (2015). *Biological autonomy: A philosophical and theoretical enquiry*. Springer.

Murat, D., Byrne, M., & Komeili, A. (2010). Cell biology of prokaryotic organelles. *Cold Spring Harbor Perspectives in Biology, 2*(10), a000422. https://doi.org/10.1101/cshperspect.a000422

Neumann, S., Hans, J. C. T., Wessels, W. I., Rijpstra, C., Sinninghe, J. S., Damsté, B. K., Jetten, M. S. M., & van Niftrik, L. (2014). Isolation and characterization of a prokaryotic cell organelle from the Anammox bacterium Kuenenia Stuttgartiensis. *Molecular Microbiology, 94*(4), 794–802. https://doi.org/10.1111/mmi.12816

Nunnari, J., & Walter, P. (1996). Regulation of organelle biogenesis. *Cell, 84*(3), 389–394. https://doi.org/10.1016/S0092-8674(00)81283-0

Obado, S. O., Glover, L., & Deitsch, K. W. (2016). The nuclear envelope and gene organization in parasitic protozoa: Specializations associated with disease. *Molecular and Biochemical Parasitology, 209*(1–2), 104–113. https://doi.org/10.1016/j.molbiopara.2016.07.008

Osada, S.-I., Ohmori, S.-y., & Taira, M. (2003). XMAN1, an inner nuclear membrane protein, antagonizes BMP signaling by interacting with Smad1 in Xenopus embryos. *Development, 130*(9), 1783–1794. https://doi.org/10.1242/dev.00401

Padeken, J., & Heun, P. (2014). Nucleolus and nuclear periphery: Velcro for heterochromatin. *Current Opinion in Cell Biology, 28*(June), 54–60. https://doi.org/10.1016/j.ceb.2014.03.001

Pfeifer, F. (2015). Haloarchaea and the formation of gas vesicles. *Life, 5*(1), 385–402. https://doi.org/10.3390/life5010385

Ramdas, N. M., & Shivashankar, G. V. (2015). Cytoskeletal control of nuclear morphology and chromatin organization. *Journal of Molecular Biology, 427*(3), 695–706. https://doi.org/10.1016/j.jmb.2014.09.008

Raschdorf, O., Forstner, Y., Kolinko, I., Uebe, R., Plitzko, J. M., & Schüler, D. (2016). Genetic and ultrastructural analysis reveals the key players and initial steps of bacterial magnetosome membrane biogenesis. *PLoS Genetics, 12*(6), e1006101. https://doi.org/10.1371/journal.pgen.1006101

Raval, P. K., Garg, S. G., & Gould, S. B. (2022). Endosymbiotic selective pressure at the origin of eukaryotic cell biology. *eLife, 11*, e81033. https://doi.org/10.7554/eLife.81033

Rout, M. P., Obado, S. O., Schenkman, S., & Field, M. C. (2017). Specialising the parasite nucleus: Pores, Lamins, chromatin, and diversity. *PLoS Pathogens, 13*(3), e1006170. https://doi.org/10.1371/journal.ppat.1006170

Sacher, M., Kim, Y.-G., Lavie, A., Byung-Ha, O., & Segev, N. (2008). The TRAPP complex: Insights into its architecture and function. *Traffic, 9*(12), 2032–2042. https://doi.org/10.1111/j.1600-0854.2008.00833.x

Sallese, M., Pulvirenti, T., & Luini, A. (2006). The physiology of membrane transport and endomembrane-based signalling. *The EMBO Journal, 25*(12), 2663–2673. https://doi.org/10.1038/sj.emboj.7601172

Santarella-Mellwig, R., Pruggnaller, S., Roos, N., Mattaj, I. W., & Devos, D. P. (2013). Three-dimensional reconstruction of bacteria with a complex endomembrane system. *PLoS Biology, 11*(5), e1001565. https://doi.org/10.1371/journal.pbio.1001565

Savage, D. F., Afonso, B., Chen, A. H., & Silver, P. A. (2010). Spatially ordered dynamics of the bacterial carbon fixation machinery. *Science, 327*(5970), 1258–1261. https://doi.org/10.1126/science.1186090

Scheffel, A., Gruska, M., Faivre, D., Linaroudis, A., Plitzko, J. M., & Schüler, D. (2006). An acidic protein aligns magnetosomes along a filamentous structure in magnetotactic bacteria. *Nature, 440*(7080), 110–114. https://doi.org/10.1038/nature04382

Schwarz, D. S., & Blower, M. D. (2016). The endoplasmic reticulum: Structure, function and response to cellular signaling. *Cellular and Molecular Life Sciences, 73*(1), 79–94. https://doi.org/10.1007/s00018-015-2052-6

Scorrano, L., De Matteis, M. A., Emr, S., Giordano, F., Hajnóczky, G., Kornmann, B., Lackner, L. L., et al. (2019). Coming together to define membrane contact sites. *Nature Communications, 10*(1), 1287. https://doi.org/10.1038/s41467-019-09253-3

Segev, N. (2011). Coordination of intracellular transport steps by GTPases. *Seminars in Cell & Developmental Biology, GTPases in Intracellular Trafficking, 22*(1), 33–38. https://doi.org/10.1016/j.semcdb.2010.11.005

Settembre, C., Fraldi, A., Medina, D. L., & Ballabio, A. (2013). Signals from the lysosome: A control centre for cellular clearance and energy metabolism. *Nature Reviews Molecular Cell Biology, 14*(5), 283–296. https://doi.org/10.1038/nrm3565

Shively, J. M. (Ed.). (2006). *Complex intracellular structures in prokaryotes* (Microbiology monographs) (Vol. 2). Springer. https://doi.org/10.1007/11497158

Short, B., & Barr, F. A. (2000). The Golgi apparatus. *Current Biology, 10*(16), R583–R585. https://doi.org/10.1016/S0960-9822(00)00644-8

Simons, K., & Sampaio, J. L. (2011). Membrane organization and lipid rafts. *Cold Spring Harbor Perspectives in Biology, 3*(10), a004697. https://doi.org/10.1101/cshperspect.a004697

Tapley, E. C., & Starr, D. A. (2013). Connecting the nucleus to the cytoskeleton by SUN-KASH bridges across the nuclear envelope. *Current Opinion in Cell Biology, 25*(1), 57–62. https://doi.org/10.1016/j.ceb.2012.10.014

Toro-Nahuelpan, M., Müller, F. D., Klumpp, S., Plitzko, J. M., Bramkamp, M., & Schüler, D. (2016). Segregation of prokaryotic magnetosomes organelles is driven by treadmilling of a dynamic actin-like MamK filament. *BMC Biology, 14*(1), 88. https://doi.org/10.1186/s12915-016-0290-1

Varela, F. J. (1979). *Principles of biological autonomy*. Elsevier.

Westrate, L. M., Lee, J. E., Prinz, W. A., & Voeltz, G. K. (2015). Form follows function: The importance of endoplasmic reticulum shape. *Annual Review of Biochemistry, 84*, 791–811. https://doi.org/10.1146/annurev-biochem-072711-163501

Wu, H., Carvalho, P., & Voeltz, G. K. (2018). Here, there, and everywhere: The importance of ER membrane contact sites. *Science, 361*(6401). https://doi.org/10.1126/science.aan5835

Xu, H., & Ren, D. (2015). Lysosomal physiology. *Annual Review of Physiology, 77*, 57–80. https://doi.org/10.1146/annurev-physiol-021014-071649

Yeates, T. O., Kerfeld, C. A., Heinhorst, S., Cannon, G. C., & Shively, J. M. (2008). Protein-based organelles in bacteria: Carboxysomes and related microcompartments. *Nature Reviews Microbiology, 6*(9), 681–691. https://doi.org/10.1038/nrmicro1913

Chapter 7
Becoming an Individual by Controlling the Movement of the Parts

Abstract This chapter examines how constraints on the motility of individual components—such as symbionts and organelles—within eukaryotic cells affect their autonomous interactive capacities and how this influences the constitutive autonomy of the entire collective association. It argues that while the motility of most symbionts and organelles is controlled by the host, this control is exercised in different ways. Symbionts may lose their motile abilities but retain the capacity to interact autonomously with the host, with interactive processes controlled internally rather than by the host. In contrast, organelles lose their autonomous motility entirely but are moved via cytoskeletal-driven mechanisms controlled by the eukaryotic cell. The chapter contends that the emergence and maintenance of collective biological identities involve stringent regulation of the motile abilities of their constituent parts. This regulation restricts, but does not necessarily eliminate, the agential capacities of individual components.

Keywords Motility · Endosymbiosis · Ectosymbiosis · Biofilms · Organelles · Functional integration

7.1 Introduction

Symbiotic associations and ecosystems are pre-eminent examples of nested organisations, as the biological members of these associations exhibit distinct genomes and specific phenotypic features. As shown in Chaps. 4 and 5, the evolutionary origin of eukaryotic cells dates to two (so far proven) endosymbiotic events: the endosymbiosis between an α-proteobacterium and the proto-eukaryotic cell is at the origin of *mitochondria*, whereas the endosymbiosis between a cyanobacterium and the proto-eukaryotic cell gave rise to *plastids*. Accordingly, eukaryogenesis is currently explained as a progressive transformation of a *nested biological organisation*

This chapter is mostly based on Militello (2019).

G. Militello, *Functional Integration*, History, Philosophy and Theory of the Life Sciences 38, https://doi.org/10.1007/978-3-031-94744-5_7

into a *functionally integrated individual* that still saves some traces of its symbiotic past (Martin et al., 2015).

The interaction among the members of a collective association is complex and includes a variety of processes ranging from metabolic fluxes to chemical signals involved in coordinated gene expression. An important, yet neglected, aspect of nested associations is the *motility of their parts*, because the motile capacities of components are severely constrained by the whole association. Since a living being can reach its nutrients in the environment and interact with its surroundings by means of motile capacities, the way in which motility is controlled and constrained affects the biological capacities not only of the parts but also of the collective association as an organised whole.

This chapter explores how the constraints imposed on the motility of the individual parts (i.e. symbionts and organelles) of a eukaryotic cell affect their autonomous interactive capacities and evaluates how this affects the constitutive autonomy of the overall collective association. Accordingly, the key question of this chapter can be stated as follows: how can a *collective identity* emerge from the control and transformation of the motility of the individual parts?

To address this issue, I will analyse how the motility of the symbionts of the eukaryotic cell is controlled by the host so as to[1] enable the self-maintenance of the whole symbiotic association. The control of motility occupies a decisive role not only in ongoing symbiotic associations but also in the transformation of endosymbiotic proto-mitochondria and proto-plastids into eukaryotic organelles: indeed, the eukaryotic cytoskeleton tightly controls the movement of eukaryotic organelles in such a way that physiological functions and homeostatic regulatory mechanisms can be performed. Accordingly, from an evolutionary point of view, the eukaryotic cytoskeleton has introduced biological novelties that permitted a proto-eukaryotic cell and its endosymbionts to achieve a functionally integrated individuality.

In the light of the above, the main issue of this chapter will be explored by addressing the following theoretical questions. How is the motility of symbionts controlled by the host so as to enable the self-maintenance of the overall symbiotic association? How is the motility of eukaryotic organelles controlled by cytoskeleton? What is the role played by the eukaryotic cytoskeleton in controlling the interactive capacities of endosymbionts and organelles and how does it affect the biological identity of the eukaryotic cell?

The analysis of these three questions sheds light on the organisational role played by motility in symbiotic associations as well as in individuals (i.e. the eukaryotic cell) based on the integration of closely related units (i.e. eukaryotic organelles). Furthermore, the different interactive behaviours of symbionts and organelles will

[1] In this chapter, I explore the relationship between motility and self-maintenance by employing some expressions (so as to, in order to, etc.) that can suggest a teleological meaning. However, all these "teleological" expressions should be understood within the organisational framework for biological functions, according to which biological functions (including motile capacities and sensorimotor abilities) are aimed at self-maintaining a biological organisation within a regime of organisational closure (see, for example, Moreno & Mossio, 2015, Chap. 3; Mossio & Bich, 2017).

shed light on their different organisational roles within the eukaryotic cell and explain why they are differently controlled.

The chapter is divided as follows: in Sect. 7.2, I present a critical review of the current debate on the individuality of symbiotic associations and some theoretical accounts of the relationship between *interactive* and *constitutive* autonomy. The following two sections will examine the physical constraints acting on the motility of eukaryotic symbionts (Sect. 7.3) and eukaryotic organelles (Sect. 7.4). Section 7.5 will explore the role played by the eukaryotic cytoskeleton in the control of motility and the evolutionary innovations that it has introduced. Finally, Sect. 7.6 makes some concluding remarks concerning the relationship between motility and biological autonomy.

7.2 Interactions at the Cornerstone of Autonomous Organisms

Over the past years, an increasing number of studies have stressed the importance of taking into consideration symbiotic interactions for defining a biological individual. The eukaryotic cell, notably in multicellular organisations, forms a nested ecosystem with their bacterial symbionts in such a way that they form a unique collective identity based on their mutual interactions (McFall-Ngai et al., 2013). Although the term "holobiont" currently designates the relationship between a multicellular eukaryote with its bacterial symbionts, Margulis (1993) employed this term to refer to a general symbiotic association between a symbiont and a host. The variety of symbiotic associations is extremely wide, since they range from prokaryote-prokaryote interactions (e.g. the *Candidatus Tremblaya princeps-Candidatus Moranella endobia* consortium of *Planococcus citri* (McCutcheon & von Dohlen, 2011) to bacterial communities of biofilms (Saxena et al., 2019), protist-prokaryote relationships (e.g. the *Paulinella chromatophora-cyanobacteria* couple) (Bodył et al., 2007), protist-multicellular eukaryotes relationships (e.g. *Giardia lamblia* and the gut of many mammals) (Adam, 2001), and prokaryotes-multicellular eukaryotes associations (e.g. the bacteria living within human gut) (Thursby & Juge, 2017). On the basis of the location of the symbiont with respect to the host, one distinguishes ectosymbionts (or epibionts) from endosymbionts (Moya et al., 2008): the former live on the surface of their host, whereas the latter within them.

All the aforementioned symbiotic associations are able to *self-maintain* by means of a number of *constitutive* interactions among symbiotic partners: metabolic, genetic, developmental and immunological interactions (Moya et al., 2008; Gilbert et al., 2012). Metabolic relationships occur when symbiotic partners interchange metabolites, nutrients and enzymes in such a way that the host provides the symbiont with the nutrients and, in turn, the symbiont supplies the host with the necessary enzymes for assimilating these nutrients or for synthesising metabolic components

(Moya et al., 2008). Genetic interactions consist in the interchange of genetic material among symbiotic partners; this phenomenon, also called as "horizontal gene transfer" (HGT), favours genetic variability and is an important source of phenotypic complexity (Ochman & Moran, 2001; Moran, 2007). The development of many invertebrates and vertebrates is partly dependent on their symbionts, because symbionts may provide larvae or embryos of the host with nutrients in such a way that "development then becomes a matter of interspecies communication" (Gilbert et al., 2012, p. 328). Finally, the immune system of the host provides its symbionts with niches where they can grow and, in turn, symbionts enhance the pathogen immunity of their host (Chiu & Gilbert, 2015; Gilbert & Tauber, 2016).

The capacity of self-maintenance of nested biological organisations needs to be studied in close connection with their ability to *interact* with the surroundings. Studies on prokaryotic endosymbionts of insects have suggested that these prokaryotes exhibit a highly reduced number of genes for cell motility (Moya et al., 2008; Degnan et al., 2010; Manzano-Marín et al., 2012). This suggests that endosymbiosis and maybe also ectosymbiosis impose some constraints on the motility of the individual parts in such a way that the motility of the symbiont(s) is modified and sometimes restricted. One of the reasons why symbiotic associations (particularly endosymbionts) exhibit different environmental conditions compared to the free-living lifestyle is that the micro-environment provided by the host generates a niche with different conditions of life compared to free-living organisms (Moya et al., 2008).

In Chap. 2, we have seen that the autonomy of a biological organisation relies on two main dimensions: the *constitutive* aspect and the *interactive* dimension. The former includes all those aspects (e.g. metabolism, regulatory processes, immunology, development, etc.) that contribute to the self-maintenance of an individual. The latter entails the capacities (e.g. perception, motility, action) that allow an organism to interact with the environment and to change it according to its own internal norms (Moreno & Mossio, 2015; Mossio & Bich, 2017).

The constitutive and the interactive dimensions are *mutually dependent*, giving rise to an "organisational closure" in such a way that the environment constrains the internal processes of an agent, and an agent exerts some constraints on its own boundary conditions (see Sect. 2.3.3). Indeed, a living being could not undergo metabolic processes, if it had not access to the nutrients that are present in the environment. Therefore, minimal forms of agency are required to allow an organism to reach its nutrients, prey or escape from its predator. In this respect, we can state that the constitutive dimension requires the interactive one. Nonetheless, the opposite holds true as well: the interactive capacities need not only the energy (e.g. in the form of ATP molecules) supplied by metabolic processes but also regulatory mechanisms that adapt agential capacities to the features of the environment. Accordingly, the interactive dimension entails the constitutive one and it could not exist without it.

The concept of "*agency*", which plays a major role both in life and cognitive sciences, summarises the main aspects of the *autonomous interactive dimension*. Indeed, an individual is an agent if it exhibits a clear distinction between the interior (e.g. the cellular environment) and the exterior (e.g. the surroundings) (individuality criterion); if it is the source of activity (interactional asymmetry criterion); and if it acts according to its own norms or goals (normativity criterion) (Barandiaran et al., 2009). An agent must be able to *modulate* and *control* its behaviour in accordance with environmental circumstances, which, in turn, is possible only if a system "is able to evaluate sequentially temporal situations and determine which possibility is functional at each moment in time. […] Thus, an agent has the ability not just to avoid negative tendencies, but to actively seek to improve its situation" (Moreno, 2018, p. 293).

In this sense, agency is a kind of *adaptive behaviour* that can be fulfilled by two different types of mechanisms: either by modifying the constitutive organisation of the system (i.e. metabolism or development) or by modifying the external conditions of the system (i.e. modification of the environmental conditions of the system). Moreno (2018) proposes a simple but valuable model for explaining an autonomous minimal agent: a system is a minimal agent if it has a regulatory subsystem that *modulates* all those inputs that produce *functional modifications* of the *environmental* conditions. The regulatory subsystem consists of a self-production network (i.e. a metabolic system) and a dynamically decoupled regulatory subsystem exerting control actions (Moreno, 2018, p. 295). Within this theoretical framework, agency is a cyclical process that requires that "the effector processes be modulated in accordance with the detected environmental conditions" (Moreno, 2018, p. 296).

A very important aspect of agency is motility, which is "an agent's capacity to move under its own power, so that it is able to perform fast (relative to its size) directional movements aimed at changing its environment in search of more favorable conditions" (Moreno & Mossio, 2015, p. 102). Motion favours a specific position of the agent with respect to its surroundings in such a way that "motility-based interaction (i.e. behavior) embeds the agent in an active sensorimotor coupling with the environment" (Arnellos & Moreno, 2015, p. 334). It has been claimed that all agents (from the simplest prokaryotes to the most complex multicellular eukaryotes) exhibit a *coupling* between *sensory inputs* (e.g. environmental cues, attractants, or repellents) and *motor capacities* in such a way that perception and action are inextricably connected (Moreno & Etxeberria, 2005; Moreno & Mossio, 2015; Di Paolo et al., 2017).[2] Agential behaviour is strongly influenced by environmental stimuli and size-time limitations[3] (Moreno & Etxeberria, 2005; Moreno & Mossio, 2015).

To conclude, the concept of "agency" has been studied in free-living organisms in close connection with their sensorimotor abilities. Nevertheless, symbiotic

[2] A clear example of sensorimotor coupling is bacterial chemotaxis (e.g. in *E. coli*), since the detection of attractants or repellents in the environment triggers a signalling cascade that modifies the frequency and the direction of the motile system (i.e. flagella).

[3] As pointed out by Moreno and Etxeberria (2005) and Barandiaran and Moreno (2008), motility and behavioural agency are strongly affected by the size of the organism, because the increase in size makes more difficult not only the correlation between sensor and effector surfaces "because of the slow velocity of diffusion processes" (Moreno & Mossio, 2015, p. 103), but also the achievement of a bodily coordination for displacement.

associations pose different constraints on the motility of their individual members in such a way that the *organisational conditions* for *agency* in nested biological associations are distinct from those of free-living organisms. This fundamental aspect of symbiotic interactions will be addressed in the following section.

7.3 The Control of Symbiotic Motility

Some prokaryotic cells are endowed with very efficient motile systems that provide them not only with the essential means of locomotion but also with an important material constraint on metabolism. Indeed, the supply of nutrients is made possible by a specific system that links the picking up of environmental cues of nutrients with locomotion. The locomotion of prokaryotes is performed by three kinds of systems: flagella, type IV pili, and cytoskeletal- and cell surface-based movements (Jarrell & McBride, 2008). Bacterial symbionts of unicellular and multicellular eukaryotes are broadly characterised by the modification of their motility systems and, more globally, interactive capacities. In this section I examine the role played by motility in the establishment of symbiotic relationships, notably I focus on three distinct symbiotic processes: *biofilms*,[4] *endosymbionts*, and *ectosymbionts*.

As explained in Chap. 5, biofilms are symbiotic communities of single- or multispecies bacteria that arise when they attach to an *abiotic* or *biotic* surface, by means of adhesins, leading to a monolayer or multilayer biofilm (Karatan & Watnick, 2009). The biofilm life cycle is characterised by important changes in the motility of its bacterial components (Fig. 7.1). At the beginning, the attachment of bacteria to a surface is strongly favoured by *flagella-mediated motility*, because flagella may facilitate the bacterial attachment to surfaces by overcoming *repulsive forces* at the surface-medium interface. Flagella may also promote the bacterial movement of growing cells along an abiotic surface in such a way that the spread of a biofilm is encouraged (Pratt & Kolter, 1998). The attachment to a surface is also promoted by type IV pili, because they contain a specific adhesin (the mannose-specific adhesin, FimH) that allows a stable cell-to-surface attachment (O'Toole & Kolter, 1998; Pratt & Kolter, 1998).

When a bacterial population increases and overcomes a size threshold, the motility of individual bacteria is *inhibited* in order to promote the constitution of the extracellular polymeric substance (EPS) matrix. The reduction of motility is

[4]Although biofilms are a kind of symbiotic association that can live independently from a eukaryotic host (indeed, biofilms can attach to abiotic surfaces), they usually attach to biotic surfaces provided by a (multicellular) eukaryotic host. Accordingly, biofilms can be considered as a specific kind of transient symbiont (i.e. a parasite) of eukaryotic cells and, therefore, it is useful to evaluate the constraints posed on the motility of the bacterial components by the extracellular polymeric matrix and how this affects the relationship with the eukaryotic host.

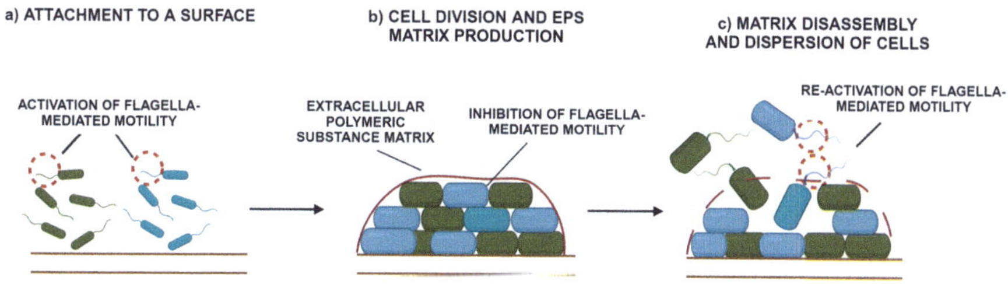

Fig. 7.1 Role of flagella-mediated motility in biofilm dynamics. (**a**) Flagella-driven motility facilitates the initial attachment of cells to a surface, a critical first step in biofilm formation. (**b**) Once attached, cell division and the production of an extracellular polymeric substance (EPS) matrix lead to biofilm maturation, during which motility is inhibited to stabilise the biofilm structure. (**c**) Upon environmental cues, motility is reactivated, enabling cells to disperse from the biofilm, thus contributing to the spread and colonisation of new surfaces

achieved by means of post-translational modifications,[5] transcriptional regulation,[6] and quorum sensing (QS) system[7] (Guttenplan & Kearns, 2013). During the existence of the EPS matrix, the motility of single bacteria is impeded. However, the EPS matrix is an ephemeral structure that disassembles in response to environmental substances concentration or bacterial lysis. The *re-activation of the genes* responsible for bacterial motility is a crucial aspect of the disassembly of the EPS matrix and, therefore, the destruction of a biofilm and the re-appearance of the planktonic state. The dispersion of a biofilm can be promoted by the synthesis of bacterial flagella (as in *Escherichia coli*) or by the production of mushroom-like pillars of bacteria (as in *Pseudomonas aeruginosa*) (Karatan & Watnick, 2009).

It is worth stressing that in biofilms the inhibition of bacterial motility is *not* performed by a (abiotic or biotic) surface, but it is rather the outcome of the signals triggered by the EPS matrix. Biofilm is an interesting case of how the *collective control of the motility of parts* allows the emergence of nested biological organisation. However, let us focus now on two kinds of symbiotic associations—endosymbiosis and ectosymbiosis—in which the motility of the symbiont is controlled by the *host*.

The *inhibition of motility* is common in bacterial endosymbionts and is due either to the *loss of the genes for cell motility* or to the *recruitment of ancient motile genes to new functions*. The loss of genes is a common aspect of intracellular bacteria and

[5] One of the most relevant post-translational modifications is the bond between the second-messenger c-di-GMP and the PilZ domain in the *ycgR* gene (Hengge, 2009).

[6] A number of transcriptional regulatory proteins may either activate (e.g. Rcs system and CsrA) or inhibit (e.g. FliZ and CsgD) the expression of flagellar genes in such a way that motility gene expression appears to be strongly controlled during the transition from motile to sessile state of bacteria.

[7] QS system plays an important role in the inhibition of chemotaxis and motion of bacteria. For example, the autoinducer 2 (AI-2) determines a cascade of events that dephosphorylate the response regulator CheY, leading to a counter-clockwise rotation of flagella and smooth swimming (Blat & Eisenbach, 1994).

parasites (Moran & Wernegreen, 2000; Gil et al., 2004), since the stable environment provided by the host and, sometimes, the existence of secondary endosymbionts make some genes redundant (Pérez-Brocal et al., 2006). In endosymbionts, the loss of genes includes both those related to metabolic processes and those associated with the synthesis of the proteins of flagellar apparatus. As a result, their motility is completely lost. A representative example is provided by *Erwinia dacicola* (a prokaryotic symbiont of the Olive Fly *Bactrocera oleae*), which has a reduced number of genes for the amino acid and carbohydrate transport and metabolism and a nearly complete loss of genes for cell motility compared to its free-living state (Estes et al., 2018).

Some endosymbionts, like *Buchnera aphidicola* (an endosymbiotic bacterium of pea aphids) (Fig. 7.2), keep their motile genes, but they cannot move, because the proteins expressed by their flagellar genes are supposed to be employed for protein transport functions, and not for motile functions (Maezawa et al., 2006). Flagellar genes are therefore used for a different purpose (likely protein transport), even though a potential pathogenic role cannot be excluded (Moya et al., 2008). As Toft and Fares (2008) pointed out, the endosymbiotic bacteria of insects usually lose their flagellar genes and they retain only the proteins of flagellum involved in protein export, whereas those involved in the synthesis of the hook and filament of flagella have generally been lost. Therefore, since the presence of flagella is unnecessary and energetically expensive, it has been suggested that the re-functionalisation of the flagellar genes of endosymbionts (like in *B. aphidicola*) is the outcome of the adaptation of the symbiont to the intracellular niche of the host (Toft & Fares, 2008).

Spirochaetes[8] live on the surface—as ectosymbionts—of many protists (within the hindgut of termites) *without performing locomotion* (Iida et al., 2000; König

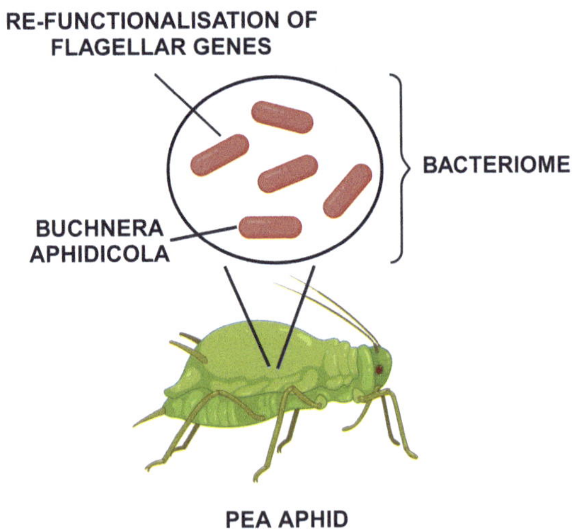

Fig. 7.2 *Buchnera aphidicola* in pea aphid bacteriome. *Buchnera aphidicola*, a symbiotic bacterium residing in the bacteriome of the pea aphid, shows re-functionalisation of flagellar genes. These genes, traditionally associated with motility, may have adapted to new roles within the symbiotic environment

[8] Spirochaetes are bacteria with spiral shapes.

et al., 2006). In spite of having flagella, spirochaetes cannot use them to move. However, the *unique* (so far known) example of bacterial ectosymbionts performing locomotion is represented by the spirochaetes living on *Myxotricha paradoxa* (a protist of the order of *Trichomonadida*) (Wenzel et al., 2003; König et al., 2006) (Fig. 7.3). *M. paradoxa* contains both endosymbionts (rod-like bacteria) and ectosymbionts (spirochaetes). Although *M. paradoxa* possesses four flagella,[9] its movement is performed by its spirochaetes. It has been proven that the loss of ectosymbionts or their inhibition by means of starvation or antibiotic treatment makes *M. paradoxa* unable to move (Radek & Nitsch, 2007). It is worth noting that many termite flagellates have been reported to have ectosymbionts with spirochaetes, but *only M. paradoxa* has spirochaetes that perform a coordinated movement in such a way that *M. paradoxa* can displace (Cleveland & Cleveland, 1966). The association of *M. paradoxa* and its ectosymbionts seems to be obligate not only for the movement but also for the performance of other vital functions of the symbiotic inter-identity (Radek & Nitsch, 2007). By contrast, the endosymbionts of *M. paradoxa*, as most of endosymbionts, cannot perform movement and are thought to perform a mitochondrion-like role.

The three symbiotic processes that I have so far examined reveal some important differences between them. Biofilms use the motility of single bacteria for the primary attaching phase; then, when the EPS matrix begins to develop, the genes for motility are inhibited. During the breakdown of the EPS matrix, the genes for motility are re-activated and they allow single bacteria to get into the planktonic state. Endosymbiosis usually promotes the inhibition of symbiont motility especially through the loss or re-functionalisation of genes for motility. Finally, ectosymbionts exhibit flagella that cannot move, except for the ectosymbiotic spirochaetes of *M. paradoxa*.

In general, in each of these three cases, the control of the motile interaction is a way to *contribute* to the *self-maintenance* of the overall symbiotic association. Indeed, the inhibition of motility of the bacteria of a biofilm keeps them in a stable position so as to favour the formation and the maintenance of the EPS matrix which in turn allows bacteria to interchange nutrients, metabolites, and to increase their

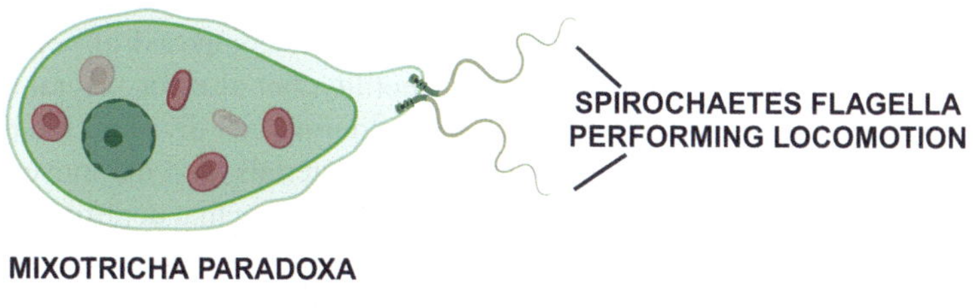

Fig. 7.3 *Mixotricha paradoxa* and spirochaetes. *Mixotricha paradoxa*, a unique protist, relies on the coordinated movement of attached spirochaete bacteria for locomotion. These spirochaetes use their flagella to propel the protist, illustrating a remarkable example of symbiosis where microbial partners perform essential functions for the host organism's mobility

[9]The flagella of *M. paradoxa* seem to be an ancient relic rather than a functional part of the protist.

immune response to pathogens and antibiotics. Likewise, the control of motility of endosymbionts and ectosymbionts indirectly affects the self-maintenance of the overall symbiotic association, because the loss of motile genes allows symbionts to spare ATP molecules that can be employed for performing physiological (notably metabolic) processes that are crucial for the whole association. Furthermore, the re-functionalisation of motile genes allows symbionts to perform important mechanisms (e.g. protein transport) that improve the metabolic relationships between the symbiont and the host. Finally, the spirochaetes of *M. paradoxa* make a direct contribution to the motility of the overall symbiotic association and as such enable it to reach its nutrients and to autonomously interact with its surroundings.

Endosymbionts arouse particular theoretical interest, as this form of symbiosis is considered as the root of eukaryogenesis, notably of mitochondria and plastids (Sagan, 1967). We may therefore suppose that the inhibition of motility, which plays a cardinal role in endosymbionts, should also be an important feature for understanding the transition from the endosymbiotic to the organelle form of mitochondria and plastids.

7.4 Mobility of Eukaryotic Organelles

Both mitochondria and plastids exhibit extremely reduced genomes and can synthesise few proteins involved in the electron transport chain and F_0F_1ATPase (mitochondria) or in the photosynthetic apparatus and in the transcription/translation apparatus (plastids). Thus, they lack almost all the genes of prokaryotic origin for the most fundamental cellular physiological functions, including those for flagella. Although neither mitochondria nor plastids can *spontaneously move*, they *are instead moved* by the *eukaryotic cytoskeleton*. Since the motility of mitochondria and plastids is *allo-driven* by cytoskeletal filaments and not self-driven by the organelle itself, they exhibit *mobility* and *not motility*. By the former, I mean the movement of an entity performed by another entity; whereas the latter is the motion performed by the entity itself.

Mitochondria and plastids *are moved* by two main cytoskeletal filaments: *microtubules* and *microfilaments*[10] (Fig. 7.4). The former are composed of *polymers of tubulin* that are responsible not only for cell motility, but also for several cellular functions, such as the transport of chromosomes during cell division, the maintenance of cell shape, the transport of intracellular materials, and the movement of cell membrane components. The latter are *filaments of actin* that control cell motility and cell separation (cytokinesis). Microfilaments can generate movement in two ways: by a sliding movement of actin and myosin filaments against each other or assembling and disassembling the microfilament bundles. In the former case, when

[10] A third system, which can be found in the eukaryotic cells of vertebrates and some invertebrates, is represented by the intermediate filaments which contribute to the maintenance of cell-shape.

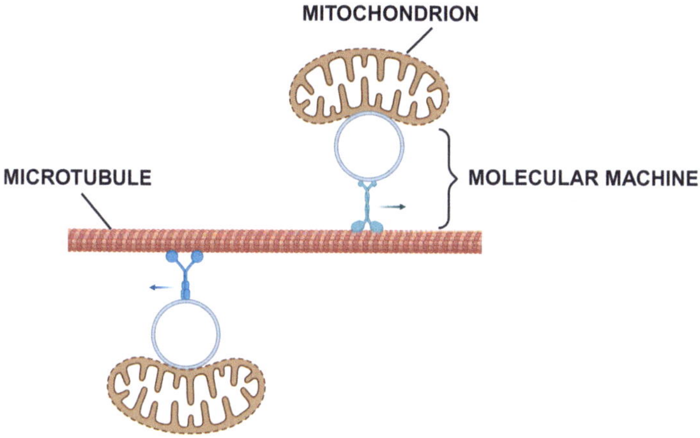

Fig. 7.4 Mitochondrial movement. The figure represents the movement of mitochondria along microtubules. This is driven by molecular motor proteins, such as kinesin and dynein, which transport the organelles to different cellular locations. The figure illustrates the interaction between mitochondria and microtubules, highlighting the role of these motor proteins in facilitating intracellular transport

myosin heads bind ATP molecules, they have a high affinity for actin and this drives the bond between actin and myosin. The hydrolysis of ATP allows myosin heads to slightly rotate and to become disengaged from myosin.[11] In the latter case, actin filaments polymerise and depolymerise so as to produce motion.

Mitochondria use cytoskeletal proteins as tracks for their *directional* (anterograde or retrograde) *movement* by means of a coordinated action between microtubules and microfilaments (Anesti & Scorrano, 2006). Both microtubules and microfilaments are important for mitochondrial movement and contribute to mitochondrial displacement in a different way. A protein (the mitochondria-microtubule binder protein, mmb1p) seems to be responsible for the bond between *mitochondria and microtubules* (Fu et al., 2011), giving rise to a functional interdependence between them. On the one hand, mitochondria reduce microtubule shrinkage rate and contribute to the stabilisation of microtubules; on the other hand, they are controlled by microtubules, because microtubules are *scaffolds* to maintain the position of mitochondria (Pon, 2011). Furthermore, the bond between *mitochondria and actin cables*, mediated by the mitochore complex, drives mitochondrial movement both in an *anterograde and a retrograde direction*. The anterograde movement of mitochondria is driven by the Arp2/3 complex[12] that stimulates actin polymerisation for the generation of anterograde force (Boldogh & Pon, 2006; Wu et al., 2013). Finally, intermediate filaments maintain cell shape by bearing tension, whereas

---

[11] In muscle cells the sliding movement is mediated by tropomyosin and troponin, which bind to the actin filament (Cappuccinelli, 1980).

[12] The Arp2/3 is a protein complex that regulates the polymerisation and depolymerisation of actin filaments.

microtubules resist compression (Wu et al., 2013). The movement of mitochondria along actin and tubulin is made possible by molecular motors[13] (myosin binds to actin, whereas dynein and kinesin bind to tubulin) which are proteins powered by ATP hydrolysis and consisting of three main parts: the *head domain* binding the cytoskeletal filament, the *neck domain* acting as a lever arm for transducing chemical energy into mechanical energy, and the *tail domain* binding the cargo. Molecular motors bind organelles at the tail domain and cytoskeletal filaments at the head domain in such a way as to act as a "cart" for the movement of organelles.

The movement of plastids is mainly due to actin filaments which are localised at the interface between these organelles and the plasma membrane (Higa et al., 2014). The motor proteins responsible for plastid movement are different from those involved in mitochondrial movement (i.e. myosin, dynein, and kinesin) and are based on the *actomyosin system* (Shimmen & Yokota, 2004). Actin polymerisation is induced by environmental stimuli (e.g. changes in light intensity or mechanical touch) and is considered the most likely candidate mechanism to generate the force required for plastids movement (see Wada & Kong, 2018).[14] Microtubules of plant cells are thought to contribute to plastid movement inasmuch as they support the functioning of actin filaments (Brandizzi & Wasteneys, 2013).

Both *mitochondrial and plastid movement* make a substantial contribution to the physiology of the eukaryotic cell, insofar as mitochondria and plastids can be more spatially close to the other eukaryotic organelles and hence favour *intracellular communication.*

Cytoskeletal-driven movement is intimately connected with the so-called "mitochondrial dynamics": the disassembly of microtubules eliminates mitochondrial mobility and permits fusion and fission events (Bartolák-Suki et al., 2017). Fusion and fission events involve changes both in mitochondrial shape and in mitochondrial membranes, as fusion entails the merger of mitochondrial membranes, whereas fission the formation of a septum within the membrane, leading to daughter mitochondria. Fusion and fission play a pivotal role in several eukaryotic cellular processes, insofar as they are involved in the maintenance of calcium homeostasis through the connection with endoplasmic reticulum, cell development and cellular division. Furthermore, mitochondrial dynamics are involved in *cell survival processes*, including autophagy, apoptosis, and necroptosis (Xie et al., 2018). The mobility of mitochondria involves not only their fusion and fission but also their capacity to interact with other eukaryotic organelles via *signalling pathways* in such a way as to regulate many cellular functions. More particularly, mitochondria interact with endoplasmic reticulum, peroxisomes, lysosomes, and Golgi apparatus (Diogo et al., 2018).

In plants, the movement of chloroplasts is important for plant growth and development. Depending on light intensity, plastids can distribute differently in the plant

[13] For a detailed discussion of molecular motors, see Militello and Moreno (2018).

[14] Although the exact mechanisms for actin polymerisation are not yet clearly understood, the protein Chloroplast Unusual Positioning 1 (CHUP1) may play a major role, because it binds to profilin which supports actin assembly (Wada & Kong, 2018).

cells (randomly in bundle sheath cells, centripetally in the vascular tissue, and centrifugally around the periphery of the bundle sheath cells) so as to favour the exchange of metabolites. Both cytoplasmic ATP levels and CO_2 diffusion are important physiological factors affecting chloroplast movement and positioning (Takagi et al., 2009). Moreover, the spatial proximity of plastids to the plasma membrane permits the maximisation of the transport of CO_2 from the intercellular airspace to the site of CO_2 fixation (the plastid stroma) and, therefore, makes photosynthesis more efficient (Takagi et al., 2009).

Despite playing a different role in the control of the movement of chloroplasts and mitochondria, both actin filaments and microtubules make a significant contribution to the positioning of the organelles within the eukaryotic cell. The molecular motors of the cytoskeleton (myosin, dynein, kinesin) are crucial to the *functional integration* of the cell, because they provide the force for the displacement of organelles. In turn, as already pointed in Chap. 6, the intracellular displacement of organelles enables *intracellular communication* and other important *physiological cellular functions* (e.g. metabolic and reproductive[15]). The controlled motion of organelles occupies a crucial organisational role that, on the one hand, makes a dramatic difference with symbiotic association, and, on the other, suggests the critical importance of the *cytoskeleton* in the *transition* from prokaryotic to eukaryotic cell.

7.5 The Contribution of Cytoskeleton to Functional Integration in Eukaryotes

The previous two sections have examined the motility of symbionts and organelles, focusing on their different functional contributions to the eukaryotic cell. In both cases the control of the motility of the parts is aimed at satisfying physiological requirements of the eukaryotic cell. However, ongoing endosymbionts and organelles of endosymbiotic origin exhibit a different control of motile capacities which can be understood partly by exploring the evolutionary innovations introduced by the eukaryotic cytoskeleton (compared to the prokaryotic one), partly by analysing the different roles played by endosymbionts and organelles within the eukaryotic cell.

Despite the discovery of bacterial homologs of actin (Bork et al., 1992), tubulin (de Boer et al., 1992; RayChaudhuri & Park, 1992; Mukherjee et al., 1993) and intermediate filaments (Margolin, 2004),[16] the eukaryotic cytoskeleton performs new functions, not present in the prokaryotic cell, which allow eukaryotes to move organelles or bacterial pathogens within themselves. Compared to the prokaryotic

[15] The relationship between mitosis and the displacement of organelles is analysed in Chap. 8.

[16] Homologs proteins for actin are FtsA, MreB, MamK, ParM, and Alf; for tubulin are FtsZ, TubZ, PhuZ, and BtubA/B; and for intermediate filaments the crescentin protein (Pilhofer & Jensen, 2013).

cytoskeleton, which is involved in the production of cell wall, the maintenance of cell shape and the support for cell division, the eukaryotic one permits a new spatial organisation of the intracellular space[17] and perform several different functions, including *intracellular transport of organelles* and *intracellular communication*. Intracellular transport of organelles is enabled by the *molecular machines* of actin filaments (myosin) and microtubules (dynein and kinesin): they bind to the organelle and to the cytoskeleton and displace the organelles along the cytoskeleton by exploiting ATP hydrolysis (Dawson & Paredez, 2013; Jékely, 2014). Since both endosymbionts (of protists and insects) and organelles are embedded in eukaryotic cells having a eukaryotic cytoskeleton, both should be moved and displaced by molecular motors along actin filaments and microtubules. Nevertheless, the fact that only organelles, and not also endosymbionts, have a cytoskeleton-driven movement is closely connected with the different functional role that organelles and endosymbionts play within the eukaryotic cell.

The movement of organelles enabled by the cytoskeleton plays an important role in *intracellular communication*: the interchange of molecules (e.g. ions, proteins, lipids, etc.) among mitochondria (and plastids), endoplasmic reticulum, Golgi apparatus, lysosomes, and nucleus would not occur if these organelles were not be *spatially close*[18] (Perico & Sparkes, 2018). In turn, the delivery and the coordinated transfer of molecules enable organelles to perform important physiological tasks that collectively contribute to the self-maintenance of the eukaryotic cell. For example, the spatial proximity between endoplasmic reticulum and Golgi apparatus allows proteins to be exchanged between these organelles. Furthermore, the closeness between mitochondria and other organelles favours the interchange of reducing equivalents and ATP molecules. Since organelle movement plays such a crucial role, the eukaryotic cell *modulates* the distribution of the organelles with *spatiotemporal accuracy* by means of changes in network and motor properties (e.g. polarisation, signalling, motor mobility, etc.) (van Bergeijk et al., 2015; Ando et al., 2015).

Unlike organelles, endosymbionts require neither displacement nor a fine-tuned dynamic spatiotemporal control from the eukaryotic cell. Indeed, endosymbionts usually provide the host with the enzymes necessary for performing catabolic or anabolic pathways (e.g. the enzymes for amino acid anabolism of sap-feeding insects) which are absent or incomplete in the host. The enzymes synthesised by endosymbionts are targeted to the plasma membrane of the host through co-translation or post-translation pathway without the need for spatial proximity to the membrane contact sites of eukaryotic organelles. For these reasons, the host does not need to consume energy to displace endosymbionts and they can be kept in an extremely stable position during the symbiotic association. It is worthy of note that

[17] The eukaryotic cytoskeleton performs some mechanisms (*filament growth*, *filament shrinkage*, and *molecular motors walking on filaments*) that provide it with the force to displace organelles and give rise to a new spatial order within the cell (see Jékely, 2014).

[18] As we have seen in Chap. 6, spatial proximity is fundamental to all those exchanges that occur through membrane contact sites (zones of apposition between two organelles). Furthermore, an efficient vesicle transport requires the action of cytoskeleton (Kamal & Goldstein, 2000).

the eukaryotic cytoskeleton can be also employed by bacterial pathogens for performing invasion strategies by exploiting actin polymerisation (Haglund & Welch, 2011; Gouin et al., 2015). Therefore, the fact that (bacterial) endosymbionts are not moved by the cytoskeleton is likely not due to a cytoskeletal limitation, but rather to the uselessness of this displacement within the eukaryotic context.[19]

The eukaryotic cytoskeleton is a fundamental step not only in the transition from prokaryotic to eukaryotic cell but also in the evolution of mitochondria and plastids from long-term stable endosymbionts to organelles. The eukaryotic cytoskeleton has given rise to an extremely dynamic and interconnected network within the eukaryotic cell that has led to complex forms of intracellular *communication* and a fine-tuned *spatiotemporal localisation* of eukaryotic organelles in such a way that the degree of cohesion and mutual dependence among the parts considerably increased. This was a very important innovation during eukaryogenesis because it opened up a more sophisticated form of *intracellular communication* (vesicular transport instead of simple diffusion) and an *effective control over the positioning of organelles*. These important biological novelties have made an important contribution to the overall functional integration of the eukaryotic cell.

Special attention should be paid to the major contribution made by the eukaryotic cytoskeleton to the transition from endosymbiotic proto-mitochondria and proto-plastids to organelles. As discussed in Chap. 4, both mitochondria and plastids have an endosymbiotic origin (α-proteobacteria were likely the ancestors of mitochondria, whereas cyanobacteria of plastids) and they transformed into organelles over millions of years (Martin et al., 2015). It has been stressed that the main events that allowed endosymbionts to become organelles were the massive transfer of genes to the eukaryotic nucleus (endosymbiotic gene transfer) and the appearance of protein import machineries in the membranes of proto-mitochondria and proto-plastids (Theissen & Martin, 2006).

It is reasonable to assume that at some point of eukaryogenesis the eukaryotic cytoskeleton must have played a pivotal role in the transformation of proto-mitochondria and proto-plastids into organelles. Indeed, given that mitochondria and plastids were endosymbionts, they lost most of their genes, including those for cell motility. It is therefore likely that in an initial phase of eukaryogenesis mitochondria and plastids were immobile or, at least, with a very reduced ability to move. However, since proto-mitochondria and proto-plastids were progressively performing regulatory and homeostatic mechanisms, it was necessary to provide some mechanisms for displacing and putting them close to other eukaryotic organelles in order to ensure intracellular communication. From this perspective, the eukaryotic cytoskeleton is no longer just a bunch of filaments for controlling cell shape, but an extremely dynamic structure that has allowed mitochondria, plastids, and the other eukaryotic organelles to achieve a high degree of functional integration.

[19] As shown in Chap. 4, endosymbionts are usually segregated in very specific parts of the cell (e.g. bacteriocytes).

7.6 Concluding Remarks

In the light of the theoretical results achieved in the previous sections, I shall explore in this concluding section how the control of the motility of the individual parts affects their *interactive autonomy* (i.e. agency) and the *constitutive autonomy* of the whole collective organisation.

The inhibition of motility is a biological phenomenon that both symbionts (except for the ectosymbionts of *M. paradoxa*) and organelles have in common. Nevertheless, I have shown that the eukaryotic cytoskeleton provides organelles with a mobility which is completely controlled by the eukaryotic cell. In the light of the distinction between mobility and motility, it is therefore clear that the notion of "motility" implies the concept of "agency", inasmuch as the *autonomous movement* is a way to interact and functionally modify the surroundings. Since both symbionts and organelles have lost their motile capacities or, if they are present, they are driven by the eukaryotic cell, is it possible to consider (endo)symbionts and organelles genuine agents?

To address this question, let us consider what a minimal agent is and then evaluate whether symbionts and organelles satisfy the conditions for minimal agency. A definition of minimal agency has been provided by Moreno (2018), who has stressed that a minimal agent is a system detecting relevant features of the surroundings (e.g. nutrients) and triggering processes that can functionally modify the environmental conditions. The effector mechanisms must be controlled *from within* by means of a *self-production network* (i.e. metabolism) and *a regulatory system* that is dynamically decoupled from the self-production network (Moreno, 2018, p. 295).

The bacteria forming a biofilm and attaching to the biotic surface of a multicellular eukaryote can detect environmental signals and nutrients that are present in the surface and perform effector mechanisms that modify their host. For example, bacteria constituting the biofilm of dental plaque can detect environmental signals such as pH or the nutrients (amino acids, proteins, glycoproteins) provided by saliva and gingival fluid and they release enzymes that produce infectious diseases (like caries or periodontitis) or inflammatory states (like gingivitis) in the host. The release of enzymes of biofilms is tightly controlled by the QS system of biofilms. Likewise, endosymbionts detect the nutrients released by their host in the host cytoplasm and they synthesise and release enzymes for metabolic pathways (e.g. the enzymes for amino acid synthesis). The production of enzymes is controlled by the genes of the endosymbiont, not by the host. Ectosymbionts (like the spirochaetes of *M. paradoxa*) detect environmental signals that activate their flagella that in turn allow *M. paradoxa* to move. The regulation of the movement of spirochaetes is made by the symbiont and not by the host. In each of these three cases, even though motility can be inhibited or lost (in bacteria of biofilms or in endosymbionts), symbionts still preserve their ability to *autonomously* interact with their host and the interactive processes are controlled *from within* and not by the host. For this reason, they can be considered as genuine agents, even if in nested hierarchical organisations of symbionts "many functions of the individuated parts are transferred to the higher collective level. These facts often lead to an ultra-simplification of certain agents (e.g. endosymbionts)" (Moreno, 2018, p. 306).

Organelles exhibit a different organisation. They perform a wide variety of functions that go far beyond metabolic contributions (like in endosymbionts) and that include regulatory and homeostatic mechanisms of the eukaryotic cell. As such, their effector mechanisms functionally change their surroundings (i.e. the eukaryotic cell) by controlling the eukaryotic cell. A clear example is provided by mitochondrial dynamics (fusion and fission) which collectively control pivotal events of the eukaryotic cell, such as apoptosis, autophagy, cell development, etc. Furthermore, the mobility of organelles, fulfilled by the cytoskeleton, allows them to efficiently communicate with one another in such a way as to perform pivotal physiological processes. Apparently, the organelles of endosymbiotic origin seem genuine agents within a "macro-agent" represented by the eukaryotic cell. However, since almost all of their genes have been transferred to the eukaryotic nucleus, the proteins controlling their functions are genetically expressed and controlled by the eukaryotic nucleus.[20] Accordingly, given that the regulation of their effector mechanisms is placed *outside the organelle*, and *not within*, they *cannot be considered genuine agents*. For example, the key proteins regulating mitochondrial fusion (Mtf1 and Mtf2, and OPA1) and fission (Drp1, Fis1, and DnmP1), despite being placed within the outer and inner mitochondrial membrane, are *expressed* and *genetically controlled* by the genes placed in the eukaryotic nucleus. The endosymbiotic gene transfer and the genetic control and expression made by the eukaryotic nucleus represent the dividing line between organelles of endosymbiotic origin and ongoing endosymbionts.

In line with the definition of "minimal agency" provided by Moreno (2018), a minimal agent has the ability of functionally modifying its surroundings by virtue of some effector mechanisms that are controlled from within. If we accept this characterisation of minimal agents, symbionts can be considered agents, even though they do not exhibit the coupling between sensory inputs and motor outputs. Sensorimotor coupling is an important aspect of agency in prokaryotic and eukaryotic forms of life (Moreno & Etxeberria, 2005; Moreno & Mossio, 2015; Di Paolo et al., 2017); however, it fails to explain why symbionts can be considered agents and why mitochondria and plastids cannot. The acknowledgement of symbionts as genuine agents allows a better characterisation of the biological status of symbiotic associations. Indeed, the identity of a symbiotic association relies on the kind of interactions (metabolic, immunological, developmental, etc.) among symbiotic partners. The *control of the motility* of the symbiont plays a very important role in the emergence of a *collective inter-identity*, insofar as it weakens the interactive capacities of the symbionts—without completely undermining them—to the benefit of the constitutive processes (metabolism, regulatory mechanisms, development, etc.) of the symbiotic association as a whole.

[20] An exception is represented by those few genes already present in mitochondrial and plastid genomes that control oxidative metabolism (in mitochondria) and photosynthesis (in plastids).

Considering symbionts as real agents is extremely important not only for explaining the emergence of collective inter-identities, but also for clarifying the difference between endosymbionts and organelles of endosymbiotic origin. The ultimate outcome of the transition from the former to the latter was the *loss* of *autonomy* and, therefore, *agential capacities*. This can be mostly attributed to the transference of genes to the host and the subsequent control of their functions by the eukaryotic cell. The reason why mitochondria and plastids are *not* agents lies in that the genes responsible for their motility were likely lost and their movement is completely controlled by the cytoplasm (and cytoskeleton). Certainly, they perform functions that change the eukaryotic cell and exhibit motor capacities driven by cytoskeleton, but the absence of an *internal* regulation of these processes do not make them agents. The interactive capacities of mitochondria and plastids can be likened to the footballers of a table football: they "kick" the little ball and they perform an action which modifies the position of the little ball; however, their movement is completely controlled by a human being who decides *when* and *how* a footballer moves so as to push the little ball towards the goal area of the opponent.

It is worth stressing that, even though a biological system has lost its *autonomous* interactive capacities, this does not necessarily imply the complete loss of interactive capacities. The case of the organelles of endosymbiotic origin is extremely clear in this respect: although organelles have lost their autonomy and their agential abilities because of a massive endosymbiotic gene transfer that has placed their genetic control in the eukaryotic nucleus, they *interact* with the other eukaryotic organelles by means of vesicle-mediated pathways and thanks to cytoskeletal proteins.

I have so far discussed the relationship between agency and interactive capacities in symbionts and organelles. I can now provide an answer to the key question of this chapter: how is the *motility of individual parts* related to the *constitutive dimension of a collective identity*? The answer lies in the fact that the control of the motility of the part is aimed at maintaining the collective identity as a whole by constraining a flux of energy and matter. As such, it keeps the nested organisation far from thermodynamic equilibrium (Mossio & Moreno, 2010; Moreno & Mossio, 2015). Both the loss or inhibition of motility (in symbionts) and the cytoskeleton-driven mobility (in organelles) are ways to contribute to the self-maintenance of the nested organisation, inasmuch as they are a fundamental support for the maintenance of other pivotal interactions (e.g. the metabolic fluxes between the part and the whole, the intracellular communication among organelles, etc.), which collectively sustain a nested organisation as a whole.

References

Adam, R. D. (2001). Biology of Giardia lamblia. *Clinical Microbiology Reviews, 14*(3), 447–475. https://doi.org/10.1128/CMR.14.3.447-475.2001

Ando, D., Korabel, N., Huang, K. C., & Gopinathan, A. (2015). Cytoskeletal network morphology regulates intracellular transport dynamics. *Biophysical Journal, 109*(8), 1574–1582. https://doi.org/10.1016/j.bpj.2015.08.034

Anesti, V., & Scorrano, L. (2006). The relationship between mitochondrial shape and function and the cytoskeleton. *Biochimica et Biophysica Acta (BBA) - Bioenergetics, 1757*(5), 692–699. https://doi.org/10.1016/j.bbabio.2006.04.013

Arnellos, A., & Moreno, A. (2015). Multicellular agency: An organizational view. *Biology & Philosophy, 30*(3), 333–357. https://doi.org/10.1007/s10539-015-9484-0

Barandiaran, X., & Moreno, A. (2008). Adaptivity: From metabolism to behavior. *Adaptive Behavior, 16*(5), 325–344. https://doi.org/10.1177/1059712308093868

Barandiaran, X., Di Paolo, E., & Rohde, M. (2009). Defining agency: Individuality, normativity, asymmetry, and spatio-temporality in action. *Adaptive Behavior, 17*(5), 367–386. https://doi.org/10.1177/1059712309343819

Bartolák-Suki, E., Imsirovic, J., Nishibori, Y., Krishnan, R., & Suki, B. (2017). Regulation of mitochondrial structure and dynamics by the cytoskeleton and mechanical factors. *International Journal of Molecular Sciences, 18*(8), 1812. https://doi.org/10.3390/ijms18081812

Blat, Y., & Eisenbach, M. (1994). Phosphorylation-dependent binding of the chemotaxis signal molecule CheY to its phosphatase, CheZ. *Biochemistry, 33*(4), 902–906. https://doi.org/10.1021/bi00170a008

Bodył, A., Mackiewicz, P., & Stiller, J. W. (2007). The intracellular cyanobacteria of Paulinella chromatophora: Endosymbionts or organelles? *Trends in Microbiology, 15*(7), 295–296. https://doi.org/10.1016/j.tim.2007.05.002

Boldogh, I. R., & Pon, L. A. (2006). Interactions of mitochondria with the actin cytoskeleton. *Biochimica et Biophysica Acta, 1763*(5–6), 450–462. https://doi.org/10.1016/j.bbamcr.2006.02.014

Bork, P., Sander, C., & Valencia, A. (1992). An ATPase domain common to prokaryotic cell cycle proteins, sugar kinases, actin, and Hsp70 heat shock proteins. *Proceedings of the National Academy of Sciences of the United States of America, 89*(16), 7290–7294. https://doi.org/10.1073/pnas.89.16.7290

Brandizzi, F., & Wasteneys, G. O. (2013). Cytoskeleton-dependent endomembrane organization in plant cells: An emerging role for microtubules. *The Plant Journal: For Cell and Molecular Biology, 75*(2), 339–349. https://doi.org/10.1111/tpj.12227

Cappuccinelli, P. (1980). *Motility of living cells*. Springer.

Chiu, L., & Gilbert, S. F. (2015). The birth of the holobiont: Multi-species birthing through mutual scaffolding and niche construction. *Biosemiotics, 8*(2), 191–210. https://doi.org/10.1007/s12304-015-9232-5

Cleveland, L. R., & Cleveland, B. T. (1966). The locomotory waves of Koruga, Deltotrichonympha and Mixotricha. *Archive Für Protistenkunde, 109*, 39–63.

Dawson, S. C., & Paredez, A. R. (2013). Alternative cytoskeletal landscapes: Cytoskeletal novelty and evolution in basal excavate Protists. *Current Opinion in Cell Biology, 25*(1), 134–141. https://doi.org/10.1016/j.ceb.2012.11.005

de Boer, P. A. J., Crossley, R. E., & Rothfield, L. I. (1992). Roles of MinC and MinD in the site-specific Septation block mediated by the MinCDE system of Escherichia coli. *Journal of Bacteriology, 174*(1), 63–70. https://doi.org/10.1128/jb.174.1.63-70.1992

Degnan, P. H., Leonardo, T. E., Cass, B. N., Hurwitz, B., Stern, D., Gibbs, R. A., Richards, S., & Moran, N. A. (2010). Dynamics of genome evolution in facultative symbionts of aphids. *Environmental Microbiology, 12*(8), 2060–2069. https://doi.org/10.1111/j.1462-2920.2009.02085.x

Diogo, C. V., Yambire, K. F., Mosquera, L. F., Tiago Branco, F., & Raimundo, N. (2018). Mitochondrial adventures at the organelle society. *Biochemical and Biophysical Research Communications, 500*(1), 87–93. https://doi.org/10.1016/j.bbrc.2017.04.124

Estes, A. M., Hearn, D. J., Agrawal, S., Pierson, E. A., Dunning, J. C., & Hotopp. (2018). Comparative genomics of the Erwinia and Enterobacter olive Fly endosymbionts. *Scientific Reports, 8*(1), 15936. https://doi.org/10.1038/s41598-018-33809-w

Fu, C., Jain, D., Costa, J., Velve-Casquillas, G., & Tran, P. T. (2011). Mmb1p binds mitochondria to dynamic microtubules. *Current Biology, 21*(17), 1431–1439. https://doi.org/10.1016/j.cub.2011.07.013

Gil, R., Silva, F. J., Peretó, J., & Moya, A. (2004). Determination of the core of a minimal bacterial gene set. *Microbiology and Molecular Biology Review, 68*(3), 518–537. https://doi.org/10.1128/MMBR.68.3.518-537.2004

Gilbert, S. F., & Tauber, A. I. (2016). Rethinking individuality: The dialectics of the holobiont. *Biology and Philosophy, 31*(6), 839–853. https://doi.org/10.1007/s10539-016-9541-3

Gilbert, S. F., Sapp, J., & Tauber, A. I. (2012). A symbiotic view of life: We have never been individuals. *The Quarterly Review of Biology, 87*(4), 325–341. https://doi.org/10.1086/668166

Gouin, E., Quereda, J.-J., & Cossart, P. (2015). Intracellular bacteria find the right motion. *Cell, 161*(2), 199–200. https://doi.org/10.1016/j.cell.2015.03.035

Guttenplan, S. B., & Kearns, D. B. (2013). Regulation of flagellar motility during biofilm formation. *FEMS Microbiology Reviews, 37*(6), 849–871. https://doi.org/10.1111/1574-6976.12018

Haglund, C. M., & Welch, M. D. (2011). Pathogens and polymers: Microbe-host interactions illuminate the cytoskeleton. *The Journal of Cell Biology, 195*(1), 7–17. https://doi.org/10.1083/jcb.201103148

Hengge, R. (2009). Principles of C-Di-GMP signalling in bacteria. *Nature Reviews Microbiology, 7*(4), 263–273. https://doi.org/10.1038/nrmicro2109

Higa, T., Suetsugu, N., Kong, S.-G., & Wada, M. (2014). Actin-dependent plastid movement is required for motive force generation in directional nuclear movement in plants. *Proceedings of the National Academy of Sciences, 111*(11), 4327–4331. https://doi.org/10.1073/pnas.1317902111

Iida, T., Ohkuma, M., Ohtoko, K., & Kudo, T. (2000). Symbiotic spirochetes in the termite hindgut: Phylogenetic identification of ectosymbiotic spirochetes of oxymonad protists. *FEMS Microbiology Ecology, 34*(1), 17–26. https://doi.org/10.1111/j.1574-6941.2000.tb00750.x

Jarrell, K. F., & McBride, M. J. (2008). The surprisingly diverse ways that prokaryotes move. *Nature Reviews Microbiology, 6*(6), 466–476. https://doi.org/10.1038/nrmicro1900

Jékely, G. (2014). Origin and evolution of the self-organizing cytoskeleton in the network of eukaryotic organelles. *Cold Spring Harbor Perspectives in Biology, 6*(9), a016030. https://doi.org/10.1101/cshperspect.a016030

Kamal, A., & Goldstein, L. S. B. (2000). Connecting vesicle transport to the cytoskeleton. *Current Opinion in Cell Biology, 12*(4), 503–508. https://doi.org/10.1016/s0955-0674(00)00123-x

Karatan, E., & Watnick, P. (2009). Signals, regulatory networks, and materials that build and break bacterial biofilms. *Microbiology and Molecular Biology Reviews, 73*(2), 310–347. https://doi.org/10.1128/MMBR.00041-08

König, H., Li, L., Wenzel, M., & Fröhlich, J. (2006). Bacterial ectosymbionts which confer motility: Mixotricha paradoxa from the intestine of the Australian termite Mastotermes darwiniensis. In J. Overmann (Ed.), *Molecular basis of symbiosis* (Vol. 41, pp. 77–96). Springer. https://doi.org/10.1007/3-540-28221-1_5

Maezawa, K., Shigenobu, S., Taniguchi, H., Kubo, T., Aizawa, S.-i., & Morioka, M. (2006). Hundreds of flagellar basal bodies cover the cell surface of the endosymbiotic bacterium Buchnera aphidicola Sp. strain APS. *Journal of Bacteriology, 188*(18), 6539–6543. https://doi.org/10.1128/JB.00561-06

Manzano-Marín, A., Lamelas, A., Moya, A., & Latorre, A. (2012). Comparative genomics of Serratia Spp.: Two paths towards endosymbiotic life. *PLoS One, 7*(10), e47274. https://doi.org/10.1371/journal.pone.0047274

Margolin, W. (2004). Bacterial shape: Concave coiled coils curve caulobacter. *Current Biology, 14*(6), R242–R244. https://doi.org/10.1016/j.cub.2004.02.057

Margulis, L. (1993). *Symbiosis in cell evolution: Microbial communities in the archean and proterozoic eons* (2nd ed.). W.H. Freeman.

Martin, W. F., Garg, S., & Zimorski, V. (2015). Endosymbiotic theories for eukaryote origin. *Philosophical Transactions of the Royal Society of London. Series B, Biological Sciences, 370*(1678), 20140330. https://doi.org/10.1098/rstb.2014.0330

McCutcheon, J. P., & von Dohlen, C. D. (2011). An interdependent metabolic patchwork in the nested symbiosis of mealybugs. *Current Biology, 21*(16), 1366–1372. https://doi.org/10.1016/j.cub.2011.06.051

McFall-Ngai, M., Hadfield, M. G., Bosch, T. C. G., Carey, H. V., Domazet-Lošo, T., Douglas, A. E., Dubilier, N., et al. (2013). Animals in a bacterial world, a new imperative for the life sciences. *Proceedings of the National Academy of Sciences, 110*(9), 3229–3236. https://doi.org/10.1073/pnas.1218525110

Militello, G. (2019). Motility control of symbionts and organelles by the eukaryotic cell: The handling of the motile capacity of individual parts forges a collective biological identity. *Frontiers in Psychology, 10*, 2080. https://doi.org/10.3389/fpsyg.2019.02080

Militello, G., & Moreno, Á. (2018). Structural and organisational conditions for being a machine. *Biology & Philosophy, 33*(5), 35. https://doi.org/10.1007/s10539-018-9645-z

Moran, N. A. (2007). Symbiosis as an adaptive process and source of phenotypic complexity. *Proceedings of the National Academy of Sciences, 104*(Suppl 1), 8627–8633. https://doi.org/10.1073/pnas.0611659104

Moran, N. A., & Wernegreen, J. J. (2000). Lifestyle evolution in symbiotic bacteria: Insights from genomics. *Trends in Ecology & Evolution, 15*(8), 321–326. https://doi.org/10.1016/s0169-5347(00)01902-9

Moreno, A. (2018). On minimal autonomous agency: Natural and artificial. *Complex Systems, 27*(3), 289–313. https://doi.org/10.25088/ComplexSystems.27.3.289

Moreno, A., & Etxeberria, A. (2005). Agency in natural and artificial systems. *Artificial Life, 11*(1–2), 161–175. https://doi.org/10.1162/1064546053278919

Moreno, A., & Mossio, M. (2015). *Biological autonomy: A philosophical and theoretical enquiry*. Springer.

Mossio, M., & Bich, L. (2017). What makes biological organisation teleological? *Synthese, 194*(4), 1089–1114. https://doi.org/10.1007/s11229-014-0594-z

Mossio, M., & Moreno, A. (2010). Organisational closure in biological organisms. *History and Philosophy of the Life Sciences, 32*(2–3), 269–288.

Moya, A., Peretó, J., Gil, R., & Latorre, A. (2008). Learning how to live together: Genomic insights into prokaryote–animal symbioses. *Nature Reviews Genetics, 9*(3), 218–229. https://doi.org/10.1038/nrg2319

Mukherjee, A., Dai, K., & Lutkenhaus, J. (1993). Escherichia coli cell division protein FtsZ is a guanine nucleotide binding protein. *Proceedings of the National Academy of Sciences of the United States of America, 90*(3), 1053–1057. https://doi.org/10.1073/pnas.90.3.1053

O'Toole, G. A., & Kolter, R. (1998). Initiation of biofilm formation in pseudomonas fluorescens WCS365 proceeds via multiple, convergent signalling pathways: A genetic analysis. *Molecular Microbiology, 28*(3), 449–461. https://doi.org/10.1046/j.1365-2958.1998.00797.x

Ochman, H., & Moran, N. A. (2001). Genes lost and genes found: Evolution of bacterial pathogenesis and symbiosis. *Science, 292*(5519), 1096–1099. https://doi.org/10.1126/science.1058543

Paolo, D., Ezequiel, T. B., & Barandiaran, X. (2017). *Sensorimotor life: An enactive proposal*. Oxford University Press.

Pérez-Brocal, V., Gil, R., Ramos, S., Lamelas, A., Postigo, M., Michelena, J. M., Silva, F. J., Moya, A., & Latorre, A. (2006). A small microbial genome: The end of a long symbiotic relationship? *Science, 314*(5797), 312–313. https://doi.org/10.1126/science.1130441

Perico, C., & Sparkes, I. (2018). Plant organelle dynamics: Cytoskeletal control and membrane contact sites. *The New Phytologist, 220*(2), 381–394. https://doi.org/10.1111/nph.15365

Pilhofer, M., & Jensen, G. J. (2013). The bacterial cytoskeleton: More than twisted filaments. *Current Opinion in Cell Biology, 25*(1), 125–133. https://doi.org/10.1016/j.ceb.2012.10.019

Pon, L. A. (2011). Organelle transport: Mitochondria hitch a ride on dynamic microtubules. *Current Biology, 21*(17), R654–R656. https://doi.org/10.1016/j.cub.2011.07.035

Pratt, L. A., & Kolter, R. (1998). Genetic analysis of Escherichia coli biofilm formation: Roles of flagella, motility, chemotaxis and type I pili. *Molecular Microbiology, 30*(2), 285–293. https://doi.org/10.1046/j.1365-2958.1998.01061.x

Radek, R., & Nitsch, G. (2007). Ectobiotic spirochetes of flagellates from the termite Mastotermes darwiniensis: Attachment and cyst formation. *European Journal of Protistology, 43*(4), 281–294. https://doi.org/10.1016/j.ejop.2007.06.004

RayChaudhuri, D., & Park, J. T. (1992). Escherichia coli cell-division gene ftsZ encodes a novel GTP-binding protein. *Nature, 359*(6392), 251–254. https://doi.org/10.1038/359251a0

Sagan, L. (1967). On the origin of mitosing cells. *Journal of Theoretical Biology, 14*(3), 255–274. https://doi.org/10.1016/0022-5193(67)90079-3

Saxena, P., Joshi, Y., Rawat, K., & Bisht, R. (2019). Biofilms: Architecture, resistance, quorum sensing and control mechanisms. *Indian Journal of Microbiology, 59*(1), 3–12. https://doi.org/10.1007/s12088-018-0757-6

Shimmen, T., & Yokota, E. (2004). Cytoplasmic streaming in plants. *Current Opinion in Cell Biology, 16*(1), 68–72. https://doi.org/10.1016/j.ceb.2003.11.009

Takagi, S., Takamatsu, H., & Sakurai-Ozato, N. (2009). Chloroplast anchoring: Its implications for the regulation of intracellular chloroplast distribution. *Journal of Experimental Botany, 60*(12), 3301–3310. https://doi.org/10.1093/jxb/erp193

Theissen, U., & Martin, W. (2006). The difference between organelles and endosymbionts. *Current Biology, 16*(24), R1016–R1017. https://doi.org/10.1016/j.cub.2006.11.020

Thursby, E., & Juge, N. (2017). Introduction to the human gut microbiota. *Biochemical Journal, 474*(11), 1823–1836. https://doi.org/10.1042/BCJ20160510

Toft, C., & Fares, M. A. (2008). The evolution of the flagellar assembly pathway in endosymbiotic bacterial genomes. *Molecular Biology and Evolution, 25*(9), 2069–2076. https://doi.org/10.1093/molbev/msn153

van Bergeijk, P., Adrian, M., Hoogenraad, C. C., & Kapitein, L. C. (2015). Optogenetic control of organelle transport and positioning. *Nature, 518*(7537), 111–114. https://doi.org/10.1038/nature14128

Wada, M., & Kong, S.-G. (2018). Actin-mediated movement of chloroplasts. *Journal of Cell Science, 131*(2), jcs210310. https://doi.org/10.1242/jcs.210310

Wenzel, M., Radek, R., Brugerolle, G., & König, H. (2003). Identification of the ectosymbiotic bacteria of Mixotricha paradoxa involved in movement symbiosis. *European Journal of Protistology, 39*(1), 11–23. https://doi.org/10.1078/0932-4739-00893

Wu, M., Kalyanasundaram, A., & Zhu, J. (2013). Structural and biomechanical basis of mitochondrial movement in eukaryotic cells. *International Journal of Nanomedicine, 8*, 4033–4042. https://doi.org/10.2147/IJN.S52132

Xie, L.-L., Shi, F., Tan, Z., Li, Y., Bode, A. M., & Cao, Y. (2018). Mitochondrial network structure homeostasis and cell death. *Cancer Science, 109*(12), 3686–3694. https://doi.org/10.1111/cas.13830

Chapter 8
Functional Integration and Reproduction

Abstract This chapter explores the role of system-level reproduction in achieving functional integration in biological individuals by analysing and comparing binary fission in bacteria with mitosis in unicellular eukaryotes. It argues that coordinated system-level reproduction necessitates the integration of reproduction, growth, and metabolism. These processes form the physiological foundation for unifying intra- and cross-generational functions, providing insight into the continuity between constitutive processes and reproductive functions. This organisational closure is crucial for establishing physiological and evolutionary individuality, offering a cohesive perspective on what defines a biological individual.

Keywords System-level reproduction · Binary fission · Mitosis · Growth · Metabolism · Functional integration

8.1 Introduction

The previous chapters have explored the various dimensions of physiological integration in both prokaryotic and eukaryotic cells, emphasising how these factors affect the constitutive and interactive autonomy of the cell. Biological autonomous organisations are characterised not only by self-maintenance and interactive capacities but also by *reproductive processes* that allow them to *transmit part of their identity* (e.g. genetic and phenotypic features) to their offspring. Although individual organisms die, the lineage they form continues to exist across multiple generations.

The persistence of a lineage depends on two key conditions. First, both unicellular and multicellular organisms must be able to reproduce in their entirety, giving rise to new biological organisations that are genotypically and phenotypically similar to the parent organism. Second, individuals within the lineage must differentially reproduce and selectively transmit their genetic and phenotypic features to their offspring, allowing for natural selection to occur. This process leads to the evolution of the lineage over successive generations.

G. Militello, *Functional Integration*, History, Philosophy and Theory of the Life Sciences 38, https://doi.org/10.1007/978-3-031-94744-5_8

The nature of reproduction and its relationship with units of selection (i.e. evolutionary individuality) has been central to a significant debate in theoretical biology and the philosophy of biology. The main issue is whether a collective reproductive system is a necessary condition for an organism to be considered a unit of selection. One perspective holds that an organism must reproduce as a whole to qualify as an evolutionary individual (Hull, 1980; Godfrey-Smith, 2009). In contrast, another view argues that evolutionary individuality does not necessarily require a collective reproductive system or the ability to reproduce as a whole (Ereshefsky & Pedroso, 2013, 2015).

A related but less explored aspect is the relationship between system-level coordinated reproduction and functional integration in biological individuals. Chapter 5 proposed the hypothesis that a certain degree of physiological integration is necessary in collective biological organisations for unitary reproduction. This chapter aims to test this hypothesis by examining the type of physiological integration required for a cell to reproduce as a whole, leading to a parent-offspring lineage. To achieve this, I analyse *binary fission* in bacteria and *mitosis*[1] in eukaryotes as case studies. In broad terms, bacterial and archaeal binary fission involves the cell splitting into two daughter cells, each with the same DNA and cellular organisation as the parent. Mitosis, a form of binary fission occurring only in eukaryotic cells, also results in two identical daughter cells but involves the formation of a spindle apparatus and a complex reorganisation of eukaryotic organelles and cytoskeleton, which is absent in prokaryotic fission. Consequently, this study focuses solely on unicellular organisms, excluding multicellular reproduction. This analysis will enhance our understanding of how proto-eukaryotic cells evolved into reproductive and evolutionary units and offer new insights into the relationship between reproduction and individuality by addressing a fundamental, often overlooked dimension of biological individuality: *functional integration*.

Thus, the research questions of this chapter can be framed as follows. What kind of physiological integration is required for a cell to reproduce as a whole? What is the relationship between physiological integration, system-level coordinated reproduction, and biological individuality? Why did the process of eukaryogenesis involve the transition from binary fission to mitosis and meiosis?

[1] Legitimately, a reader could ask why I do not examine the other fundamental mechanism of eukaryotic reproduction, namely meiosis. Two primary reasons justify this choice. Firstly, mitosis and meiosis share many important mechanisms (e.g. regulatory proteins for the coordination of different phases of cell division or cytoskeletal proteins for the spatial organisation of the cell during division). Thus, the kind of physiological integration required for meiosis is not significantly different from that of mitosis. Secondly, mitosis is a ubiquitous feature of eukaryotes, whereas meiosis is a facultative mode of division absent in some unicellular eukaryotes (e.g. *Giardia lamblia* can exchange chromosomes without a true meiosis) (Carpenter et al., 2012). As such, the kind of functional integration required for mitosis can be applied to all eukaryotes, whereas that underlying meiosis cannot be generalised to all eukaryotic species. Nevertheless, meiosis will also be addressed in some parts of the chapter, especially in relation to mitosis.

This chapter argues that system-level coordinated reproduction requires functional integration with developmental[2] and metabolic processes. This integration results from three fundamental mechanisms that enable organisational closure between reproduction, growth, and metabolism. These insights have significant implications for understanding the relationship between physiological and evolutionary individuality.

The chapter is structured as follows: Sect. 8.2 critically reviews the ongoing debate on the relationship between reproduction and biological individuality, considering both physiological and evolutionary aspects. Sections 8.3 and 8.4 analyse the physiological integration necessary for cell division in prokaryotic and eukaryotic cells, respectively. Section 8.5 explores the interplay between functional integration, reproduction, and biological individuality through case studies. Finally, Sect. 8.6 offers concluding remarks.

8.2 The Philosophical and Biological Debate on Reproduction and Individuality

Over the last five decades, the biological and philosophical debate has linked the question of reproduction with two fundamental dimensions of biological individuality: the *evolutionary* and the *physiological*. An evolutionary individual must possess reproductive capacities that allow it to generate offspring capable of undergoing variation. At the same time, reproduction is integral to the life cycle and developmental processes of a physiological individual. Thus, reproduction appears to occupy a unique, perhaps ambiguous, status that lies between the ontogenetic and phylogenetic dimensions of biological organisation. This section aims to critically review the main perspectives on reproduction, with an emphasis on its relationship with the evolutionary and physiological dimensions of biological individuality.

The issue of the reproducibility of a biological system was first posed by Dawkins (1976, 1982) and Hull (1980), who defined a "*replicator*" as any entity capable of transmitting its biological features to its descendants. Neither Dawkins nor Hull conceptualised reproduction as such, but rather, it is implicitly invoked as a matter of replication (understood as copying) of a biological organisation, which adheres to the principles of fecundity, fidelity, and longevity (Dawkins, 1976, 1982; Hull, 1980). According to Dawkins, replicators produce a high number of copies (fecundity) and transmit (genetic) information with high fidelity during replication, thus preserving genetic information over time (Dawkins, 1982). Genes are the best candidates to be considered replicators because they can transmit genetic and phenotypic aspects of the overall organisation to new systems. Both Dawkins (Dawkins, 1976, 1982) and Hull (1980) emphasise that replication cannot occur independently

[2] In this chapter, when I employ the words "development" or "developmental processes", I use them synonymously with "growth" and vice versa.

and must be assisted by vehicles (Dawkins) or interactors[3] (Hull), which are produced by replicators and help them increase in numbers by interacting effectively with their environments (Dawkins, 1982).

In recent years, there has been a shift in focus from replication to *reproduction* because the replication of the basic units of life (i.e. cells) takes the form of reproduction. Although related, replication and reproduction are distinct concepts: the former refers to a mere resemblance between the generator and the generated and, in the case of genetic replication, to the transmission of genetic information from the replicator to the replicated. Reproduction, on the other hand, refers to biological processes (e.g. fission and fusion) characterised by a *material continuity* between parents and offspring (Griesemer, 2000, 2016). Thus, Griesemer (2000) proposed using the term "*reproducer*" instead of "replicator" to designate all entities that can multiply and transmit their material structure to their offspring.

Reproduction is central to the *Darwinian* view of natural evolution through natural selection because heritability and fitness require that organisms transmit their traits to their offspring (Godfrey-Smith, 2009). For this reason, Godfrey-Smith (2009, 2013) has argued that individuals with evolutionary capacities (i.e. evolutionary or Darwinian individuals) must include the ability to reproduce, giving rise to parent-offspring lineages. Godfrey-Smith has expanded on Griesemer's concept of reproducers and suggested that it is possible to distinguish three categories of reproducers with evolutionary capacities: simple, collective, and scaffolded reproducers. Simple reproducers (e.g. unicellular organisms) produce new entities (other cells), the components of which (i.e. cell components) cannot reproduce. Collective reproducers (e.g. multicellular organisms or symbiotic relationships) generate new organisms, the components of which (i.e. germ cells) can reproduce. Scaffolded reproducers (e.g. viruses) reproduce as part of the reproduction of some larger units (Godfrey-Smith, 2009, 2013).

Unlike the concept of a replicator, the term "reproducer" is intimately connected with the life cycles and developmental processes of a biological organisation, as development is a process of growth and differentiation of parts along with the maturation necessary for reproduction. Indeed, the life cycle (or biological cycle) of an organism includes "a series of developmental transformations and reproductive phases that lead from a given developmental stage of a given organisational form to the same developmental stage of the same organisational form in a following generation, through all the organisational forms of the organism" (Fusco & Minelli, 2019, p. 23). According to Griesemer (2016), development is not merely a life cycle phase preceding reproduction; rather, they are two mutually embedding and entwined aspects of life. The connection between reproduction and development is facilitated by "*scaffolds*", which are temporary structures that help developing entities undergo developmental (e.g. assembly and construction) and reproductive processes (Griesemer, 2016, p. 806). Furthermore, reproduction is characterised by the

[3]There is a slight difference between Dawkins' vehicles and Hull's interactors: the former are considered passively, whereas the latter are seen in a more active sense.

material overlap of parts (notably genetic and phenotypic traits) that convey or confer developmental capacities on offspring via the transfer of material components (Griesemer, 2000).

The characterisation of a life cycle is based on the idea that it is possible to sharply distinguish between *reproduction*, which entails the transition to a new generation, and *development*, which is a transformation of the same individual (Fusco & Minelli, 2019, p. 23). Nevertheless, the relationship between development and reproduction is somewhat problematic, as it is often difficult to differentiate "the transformations of an individual that do not alter its identity, and those that, at some point, will result instead in a new distinct individual, somehow emerging from it" (Fusco & Minelli, 2019, p. 25). Moreover, in many biological organisations, life cycles are complex[4] because they "involve relatively discontinuous and substantial changes of form, behaviour, or environment" (Griesemer, 2016, pp. 803–804), thus complicating the relationships between reproduction and development.

A further problem posed by reproduction lies in the persistence of biological identities across the generations of the same lineage. While it is true that reproduction entails a material overlap between parents and offspring (Griesemer, 2000, 2016), it is also true that reproduction is characterised by an *organisational discontinuity*[5]: "a functional change (an alteration, disappearance, or appearance of one or more functional constraints) occurring in the temporal unfolding of constraint dependencies" (Mossio & Pontarotti, 2022). This organisational discontinuity is the outcome of three different types of reproductive processes: fission, fusion, and sexual reproduction. Reproduction may involve the cessation of parental existence upon reproduction. This occurs in fission events such as cell division and also in fusion events. Fission entails the division of a system (e.g. a cell) into two or more parts, such that the previous system ceases to exist and two new individuals appear. Fusion events lead to a spatiotemporal separation (asymmetrical dependence relation) between the fusing systems once they have come together in space. However, there are also forms of reproduction in which parents persist after reproduction: this is the case with sexual reproduction, which entails both fusion and fission events of gametes from parents, followed by the integration of the zygote with the mother. The zygote exhibits a spatiotemporal separation from the mother from both an evolutionary and developmental point of view (Difrisco & Mossio, 2020).

Although reproduction results in an organisational change in the transition from parents to offspring, the conservation of genetic and phenotypic traits among individuals of the same lineage implies that a part of the biological organisation of the

[4] Fusco and Minelli (2019) distinguish between two broad categories of life cycles: first, monogenerational life cycles, in which there is one generation, one kind of organisation, one developmental process, and one reproductive phase; secondly, multigenerational life cycles, in which multiple generations occur, characterised by multiple organisational forms, multiple developmental processes, and multiple reproductive phases.

[5] As pointed out by Difrisco and Mossio (2020), organisational continuity (often referred to as "diachronic identity") is characteristic of the developmental stages of the same individual, presupposing a spatiotemporal continuity between the material and functional states of an organism.

parent(s) is preserved in the offspring. This phenomenon is often referred to as biological heredity, which has typically been understood as synonymous with "genetic heredity" during the twentieth century. Nevertheless, the expression "biological heredity" can be used in a more extensive way by designating the cross-generation conservation of functional elements, which are defined as constraints subject to cross-generation closure[6] (Mossio & Pontarotti, 2022). Heredity refers to the specific way in which the functional constituents of a biological system remain stable over generations ("cross-generation stability"), in contrast to the stability of various elements in the environment. Without cross-generation similarities, the discontinuity between successive generations would be interpreted as the production of entirely new and different organised systems rather than reproduction (Mossio & Pontarotti, 2022). Thus, reproduction contributes to cross-generation stability because (a) it allows a biological organisation to transmit its functions to the offspring, and (b) the hereditary object is subject to a cross-generation closure. This enables organisational stability across generations (Saborido et al., 2011).

In light of the above, it is apparent that reproduction is a fundamental link between physiological life cycles (including growth) and evolutionary individuality (i.e. transmission and variation of traits in offspring). Typically, evolutionary individuals possess system-level coordinated capacities, culminating in parent-offspring lineages. Thus, a basic question, though often neglected in the current literature on biological reproduction, is what type and degree of functional integration is required for the developmental processes of the life cycles of a biological organisation to give rise to system-level reproduction, resulting in a parent-offspring lineage. To study this issue, the next two sections will examine the types of functional integration required for system-level reproduction in prokaryotic and eukaryotic cells.

8.3 Binary Fission: How Bacterial Reproduction Is Functionally Integrated with Metabolic and Developmental Processes

Prokaryotic organisms, such as bacteria and archaea, reproduce through binary fission, a form of asexual reproduction characterised by DNA replication, chromosome segregation, and the formation of a cell wall (septum). This process enables prokaryotes to divide into two genetically identical daughter cells. In this section, I will focus on the mechanisms that integrate binary fission with development and

[6] By "cross-generation closure", the authors mean that the biological functions of an organism depend on those of the parent(s). The connection is ensured by reproductive functions (cross-generation functions) that reproduce the same biological functions of the parents in the biological organisation of the offspring.

metabolism, and how these mechanisms coordinate reproduction at a systemic level. My analysis will be limited to bacterial fission and will not extend to archaea.[7]

Before examining the interplay between reproduction, development, and metabolism, it is important to discuss the organisation of the bacterial life cycle, which is central to this section and highlights the significant continuity between developmental and reproductive processes. The bacterial cell cycle is divided into three phases: the *birth* phase (B), the *chromosome* phase (C), and the *division* phase (D). The B phase, occurring between cell birth and chromosome replication, is marked by bacterial growth in response to nutrient availability. The C phase serves as a transition between the developmental and reproductive phases, characterised by chromosome segregation and DNA replication. Finally, in the D phase, bacteria divide into two daughter cells, each with a complete copy of the parent genome (Dewachter et al., 2018) (see Fig. 8.1).

The growth and reproduction of a bacterium during phase B critically depend on the availability of nutrients in the environment. Multiple signalling pathways transmit information about nutrient levels and growth rates to the cell cycle machinery, allowing cells to adapt to fluctuations in nutrient availability (Wang & Levin, 2009). When bacteria detect sufficient nutrients, they increase in size and mass. In the absence of essential nutrients, however, they are unable to grow, leading to reduced size and mass. For instance, carbon availability and its metabolic processing are key determinants of bacterial cell size: under carbon-rich conditions, a regulatory

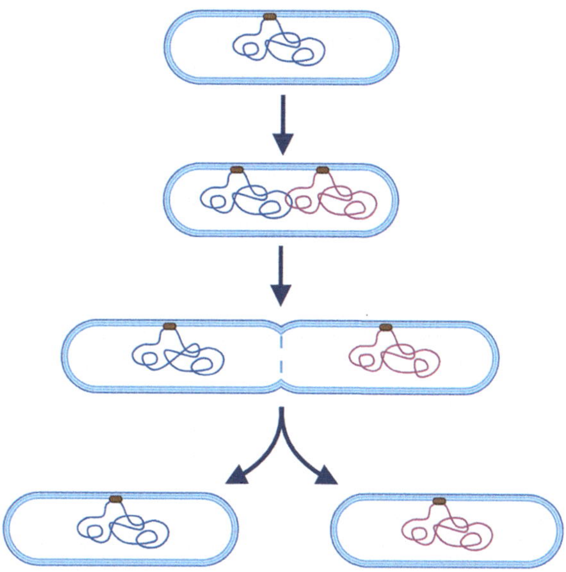

Fig. 8.1 Schematic representation of binary fission in prokaryotes. The figure depicts the process of cell division in prokaryotic cells, including the replication of DNA, elongation of the cell, and separation into two genetically identical daughter cells

[7] The main mechanisms underlying binary fission in bacteria exhibit many similarities to those of archaea. Accordingly, for simplicity, I focus on bacteria, taking them as a paradigmatic case of prokaryotic reproduction.

network stimulates an increase in cell size (Wang & Levin, 2009). This dependency on nutrient availability is crucial for maintaining average cell size under specific growth conditions.

Nutrient availability and metabolic control also influence the C and D phases of the bacterial cell cycle. When nutrients are plentiful, both DNA replication and chromosome segregation (phase C) are activated. Metabolic substrates initiate and elongate DNA replication, which in turn triggers chromosome segregation. Conversely, nutrient scarcity inhibits both replication initiation and chromosome segregation. For example, in *Escherichia coli* (a Gram-negative bacterium), amino acid starvation leads to the production of guanosine tetraphosphate and pentaphosphate,[8] which regulate the nutrient-dependent control of DnaA[9] expression and replication initiation (Wang & Levin, 2009). Similarly, in Gram-positive bacteria like *Bacillus subtilis*, sugar metabolic signals appear to control DNA replication and chromosome segregation.

Nutrients and metabolites also regulate the D phase. For example, in both *E. coli* and *B. subtilis*, the accumulation of UDP-glucose (a glycogen precursor) inhibits cell division by interfering with the activity of FtsZ, a cytoskeletal-like protein that controls binary fission. When UDP-glucose levels are low, cell division proceeds (Wang & Levin, 2009).

A key aspect of the integration between bacterial reproduction and development is their *temporal coordination*, which is mediated by a set of regulatory proteins that synchronise bacterial growth with chromosome segregation, DNA replication, and bacterial fission. This temporal coordination involves two main elements.

Firstly, DNA replication and binary fission occur only after bacteria have reached a critical cell size, typically about double their original mass, under given growth conditions (Wang & Levin, 2009; Willis & Huang, 2017). While the specific proteins responsible for coordinating bacterial growth, DNA replication, and cell division remain largely unidentified, it is believed that key protein regulators—such as the MreB protein for growth, the DnaA protein for DNA replication, and the FtsZ protein for cell division—interact to orchestrate cell cycle events and growth (Willis & Huang, 2017).

Secondly, the coordination of DNA replication and chromosome segregation (phase C) with bacterial reproduction (phase D) ensures that each cell division is accompanied by one complete round of chromosome replication (Westfall & Levin, 2017). For instance, the temporal regulation between DNA replication and binary fission is controlled by the accumulation and degradation of the active DNA-ATP complex throughout the cell cycle. Similarly, the coordination of chromosome segregation with binary fission relies on two mechanisms—nucleoid occlusion[10] and

[8] Guanosine tetraphosphate and pentaphosphate are molecules that inhibit RNA synthesis when there is a low concentration of amino acids.

[9] DnaA is a transcription factor that regulates its own expression and that of several other genes.

[10] Nucleoid occlusion is a negative regulatory system that allows the Z-ring to assemble after chromosome segregation, preventing chromosome fragmentation (Dewachter et al., 2018).

Ter linkage[11]—that facilitate the formation of the Z ring[12] after chromosome segregation (Dewachter et al., 2018).

Cytoskeletal-like proteins, such as the FtsZ protein and the MinCDE system, are crucial for the system-level coordination of bacterial reproduction. These proteins are involved in forming the Z-ring and the septum, which are essential for dividing the bacterial cytoplasm and generating two daughter cells (Wallden et al., 2016; Dewachter et al., 2018). The FtsZ protein, in particular, plays a prominent role. Homologous to eukaryotic tubulin, FtsZ assembles into a ring structure known as the Z-ring at midcell. The Z-ring acts as a dynamic, polymeric scaffold that recruits divisome components, constrains liposomes, and determines the site of septum synthesis (Dewachter et al., 2018). Additionally, FtsZ can bind GTP and convert it to GDP, providing an energetic source that may drive the contraction of the Z-ring and facilitate cell division (Margolin, 2005; Lan et al., 2009).

From this, several conclusions can be drawn about the conditions enabling system-level reproduction in bacteria. Bacterial growth and reproduction are influenced by nutritional cues and overall metabolic status. Developmental processes and reproduction are tightly coordinated to ensure that cell division is associated with reaching a specific cell size and that each cell division includes one round of chromosome replication. Cytoskeletal-like proteins, particularly FtsZ, are integral to forming the septum and Z-ring, providing the necessary machinery for bacterial division.

8.4 Functional Integration Between Mitosis, Growth, and Metabolism in Eukaryotic Cells

This section explores the physiological integration required for a eukaryotic cell to achieve system-level reproduction. I focus on mitosis—the most representative example of eukaryotic reproduction—and examine three organisational dimensions related to its integration: first, the dependence of mitosis on the cell's growth processes; secondly, the functional connection between mitosis and the spatial reorganisation of the eukaryotic organelles and cellular membrane; and thirdly, the functional integration between mitosis, growth, and metabolism.

The eukaryotic cell cycle is divided into two main parts: the *interphase*, during which the cell grows and replicates its DNA, and the *mitotic phase*, in which the cell divides into two daughter cells with identical genetic and phenotypic traits. The interphase is typically subdivided into three phases: the G1 phase, during which the cell increases in size and synthesises mRNAs and proteins; the S phase, where DNA

---

[11] Ter linkage is a protein complex (consisting of the proteins MatP, ZapA, and ZapB) that provides spatiotemporal coordination between chromosome segregation and Z-ring assembly by coupling the chromosomal terminus region to the divisome (Dewachter et al., 2018).

[12] The Z-ring is a polymer structure involved in binary fission.

replication occurs; and the G2 phase, during which the cell continues to grow and synthesises proteins. The mitotic and cytokinesis phases (the division of the cytoplasm into two daughter cells) occur after the G2 phase. Between these phases and the G1 phase, there is a quiescence phase known as the G0 phase.

Mitosis consists of five phases: *prophase*, *prometaphase*, *metaphase*, *anaphase*, and *telophase* (see Fig. 8.2). Prophase is characterised by chromosome condensation and the movement of centrosomes and microtubules to opposite poles of the cell. The mitotic spindle forms during prometaphase and attaches to the chromosomes. In metaphase, the mitotic spindle organises into a twofold symmetric structure. Anaphase is marked by the segregation of chromosomes, leading to the formation of two fully functional nuclei at opposite ends of the cell in the process known as telophase (McIntosh, 2016).

We can now examine how mitosis is integrated with the developmental processes occurring during the G1 and G2 phases of the cell cycle. Mitosis is spatiotemporally coordinated with the G2 phase (preceding mitosis) and the G1 phase (following mitosis) through the action of regulatory proteins, notably the cyclin-dependent kinases (CdKs), which orchestrate cell cycle events to ensure sequential progression. For instance, the transition from G2 to the M phase is facilitated by the Cdk1/cyclinB1 complexes. These CDK complexes, together with Polo-like kinase 1 (Plk1), inhibit the kinase Wee1 and activate the Cdc25 phosphatase, thereby triggering mitosis. During mitosis, chromosomes attach to the mitotic spindle and align at the metaphase plate, which activates the anaphase promoting complex/cyclosome

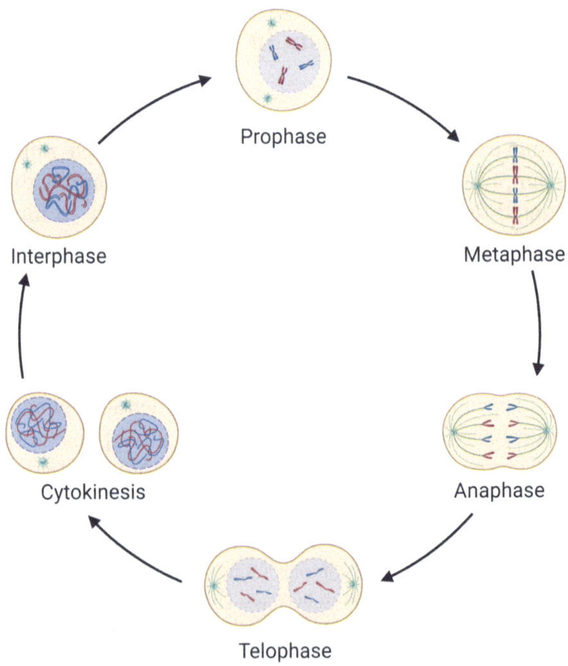

Fig. 8.2 Schematic representation of mitosis. The figure illustrates the key stages of mitosis, including prophase, metaphase, anaphase, and telophase, and shows the alignment and separation of chromosomes, as well as the formation of the mitotic spindle and the division of the nucleus into two daughter nuclei

(APC/C). This protein complex drives the exit from mitosis and entry into the G1 phase, ensuring proper chromosome segregation (Ovejero et al., 2020).

Another important aspect of mitosis is the *controlled duplication and segregation of intracellular organelles* and their transmission from parent to offspring. This requires the functional and spatiotemporal coordination of the main components of the eukaryotic cell: the action of cytoskeletal filaments, modification of the nuclear membrane, reshaping of the plasma membrane, and reshaping/disassembly of organelles. These events are coordinated not only with each other but also with the other stages of interphase, providing a functionally integrated organisation for mitosis.

The formation of the mitotic spindle[13] is enabled by *microtubules* and *tubulins*. Microtubules play a fundamental role in the regulation of mitosis because they control the shape of the mitotic spindle and facilitate the alignment of chromosomes at the spindle equator. The connection between microtubules and chromosomes is mediated by *kinetochores*, which are protein complexes located on each chromatid. These complexes include several fibrous proteins that bind to the walls of microtubules.

Kinetochores rely on molecular machines,[14] particularly motor proteins[15] such as dynein and kinesin, which bind to microtubules and generate forces that can alter both chromosome position and microtubule dynamics. All kinetochores bind minus-end-directed motors and are actively motile on the microtubules to which they attach (Wieser & Pines, 2015). During telophase, microtubules and their associated proteins are involved in assembling the cytokinetic machinery, which divides the cytoplasm into two daughter cells. Subsequently, eukaryotic cells use microfilaments and actin to complete the process of cell division (Wieser & Pines, 2015; D'Avino et al., 2015).

All the stages of mitosis involve a fundamental reshaping of the *nuclear envelope* to allow microtubules to reach chromosomes and ensure their accurate segregation.[16] This reshaping may or may not involve the breakdown of the nuclear envelope: the former is known as open mitosis, and the latter as closed mitosis. Open mitosis occurs in multicellular eukaryotes and is characterised by the disassembly of the nuclear envelope, resulting in the fusion of the nucleoplasm and the cytoplasm. Chromosomes condense and attach to spindle microtubules within a nuclear envelope-like structure, and the nuclear envelope disassembles and becomes incorporated into the endoplasmic reticulum (Boettcher & Barral, 2013). In contrast, closed mitosis occurs in unicellular eukaryotes, such as yeasts and protozoa, where

---

[13] The mitotic spindle organises DNA such that one copy of each chromosome attaches to each end of the spindle. Then, the movement of the spindle separates the duplicated chromosomes into two distinct sets and pushes them towards the opposite ends of the cell.

[14] See Militello and Moreno (2018) for a broader discussion of molecular machines.

[15] The motor proteins of the cytoskeleton have also been discussed in Chaps. 6 and 7.

[16] As shown in Sect. 6.4, cytoskeletal filaments interact with the nuclear envelope to control its position in the cell, promote the spatial connection and communication between the nuclear and the plasma membrane, and enable mitosis.

the nuclear envelope remains intact. The mitotic spindle forms in the cytoplasm and interacts with chromosomes through the nuclear envelope. Each kinetochore attaches to the inner surface of the nuclear envelope and creates a microtubule attachment site on the envelope's cytoplasmic face (McIntosh, 2016). The nucleus undergoes fission, where nuclear components (e.g. transcription factors and non-chromosomal DNA) are compartmentalised and segregated to ensure they are evenly distributed between the nuclei of the daughter cells (Boettcher & Barral, 2013).

During mitosis, the *cell membrane* undergoes significant reshaping due to changes in the surface-to-volume ratio. At the beginning of mitosis (prophase), the surface area of the cell membrane decreases because of the interruption of membrane recycling from endosomal compartments. By the end of mitosis (anaphase), the surface area increases through the extensive fusion of endosomal membranes. This synchronous and coordinated fusion of the plasma membrane with the endomembranes, facilitated by Ca^{2+} signalling pathways, allows endomembranes to be stored within the new cells (Boucrot & Kirchhausen, 2007).

The *secretory pathway* undergoes significant changes during mitosis, beginning with a general cessation of membrane traffic, including endocytosis and endosome fusion. During mitosis, endosomes and lysosomes remain intact and separate. Their segregation into daughter cells requires coordinated movements, and during cytokinesis, these organelles accumulate near the microtubules (Bergeland et al., 2001). The mechanisms underlying organelle fragmentation involve a temporary inhibition of fusion machinery and the continuous synthesis of transport vesicles. For example, both the endoplasmic reticulum and the Golgi apparatus break down from single-copy organelles into several vesicle clusters. However, endosomes and lysosomes are maintained during cell division and do not fragment or fuse, instead being partitioned as separate, intact vesicles (Bergeland et al., 2001).[17] Mitochondria and plastids divide and segregate into the two daughter cells in a process driven by cytoskeletal proteins, thus ensuring the transmission of mitochondrial and plastid DNA to the daughter cells (Imoto et al., 2011).

The *spatiotemporal coordination* between cytoskeletal activities and the reshaping of nuclear, plasma, and endomembranes is primarily controlled by cyclin-dependent kinases, ensuring coordination among the distinct phases of mitosis, including mitotic onset, sister chromatid separation, and mitotic exit. For example, the binding of cyclin B to Cdk1[18] activates key mitotic events such as nuclear envelope breakdown, chromosome condensation, mitotic cell rounding,[19] APC/C

[17] According to Bergeland et al. (2001), there is no specific mechanism for the exact distribution of endosomes to daughter cells. This process is stochastic in nature, though the endosomal compartments are clustered by directional movements, suggesting that microtubules are involved in the process.

[18] In animals, cyclin B-Cdk1 shuttles between the nucleus and the cytoplasm during interphase and then accumulates in the nucleus at prophase. This promotes the breakdown of the nuclear envelope and the disassembly of the nuclear lamina (Gavet & Pines, 2010).

[19] Mitotic cell rounding is a shape change that occurs during mitosis in most animal cells.

activation,[20] SAC signalling,[21] and kinetochore assembly, while inhibiting others such as transcription, translation, intracellular trafficking, Golgi integrity, cytokinesis, and potentially ER reorganisation and endocytosis (Wieser & Pines, 2015). The inactivation of the cyclin B1-Cdk1 complex triggers the final events of mitosis, including mitotic exit and cytokinesis, resulting in the formation of two genetically identical daughter cells.

Mitosis is coordinated not only with cellular growth but also with *metabolism*, so that the cell cycle regulates metabolism to meet the specific requirements of mitosis. In yeast, for example, the cyclin-dependent kinase (Cdk1) regulates both the cell cycle and carbon metabolism: at the G1/S transition, Cdk1 phosphorylates and activates the trehalase Ntk1; during the S/G2/M phases, the active Ntk1 releases trehalose to fuel glycolysis (Ewald et al., 2016). Thus, the mechanisms underlying cell cycle progression are coordinated with nutrient-sensing kinases, integrating metabolism with cell cycle progression. At the same time, it has been shown that metabolism regulates cell cycle progression. In mammals, some enzymes involved in glucose metabolism (e.g. 6-phosphofructo-2-kinase/fructose-2,6-bisphosphatase 3) control the passage from G1 to S and from S to G2-M. The presence of substrates for glycolysis allows for the transition from one phase to another, whereas the reduction of glucose availability impairs this progression (Kalucka et al., 2015).

To conclude, *three levels of functional integration* are required for mitosis. First, there is the spatiotemporal coordination between the stages of the *interphase* and the *mitotic phase*, which establishes a level of functional interdependence between cellular growth and division. Secondly, there is the spatiotemporal coordination between the *cytoskeleton* and the modifications of the *intracellular organelles*. In both cases, this coordination is ensured by regulatory proteins, notably cyclin-dependent kinases (CDKs), which act as cell cycle activators and inhibitors. More specifically, they (a) activate some processes of the cell cycle, such as DNA replication, by inhibiting others, like cell division; and (b) trigger certain morphological transformations and functions of cellular structures, such as chromosome condensation, by repressing others, like the inhibition of transcription and translation during mitosis. Finally, there is the coordination of *metabolism*, *growth*, and *division*, which involves both the regulation of the cell cycle through metabolites acting as signals and the regulation of metabolism by the cell cycle regulators, such as CDKs.

[20] APC/C plays a pivotal role in regulating eukaryotic cell reproduction. Ubiquitin-mediated proteolysis regulates sister chromatid separation by altering the local balance of protein kinases and phosphatases and by activating separase, a protease that cleaves the cohesion rings holding sister chromatids (Wieser & Pines, 2015).

[21] SAC stands for spindle assembly checkpoint, a set of mechanisms that control mitosis and meiosis, preventing the separation of the duplicated chromosomes (anaphase) before their proper attachment to the mitotic spindle.

8.5 Functional Integration, Reproduction, and Biological Individuality

In the previous sections, I examined the mechanisms that enable the physiological integration of cell metabolism, growth, and division. This section has a twofold purpose: first, it summarises and compares the levels of mechanistic integration in prokaryotes and eukaryotes to highlight the similarities and differences in how reproduction, growth, and metabolism are interconnected; secondly, it evaluates the philosophical implications of the interdependence between metabolism, growth, and reproduction for understanding issues of biological autonomy and individuality.

The first level of mechanistic integration is represented by *nutrient-dependent signals* (e.g. metabolites) and regulatory proteins that trigger cell growth (e.g. increase in cell size or DNA replication) and division in response to nutrient availability. In bacteria, the intermediates of glucose metabolism and regulatory proteins (e.g. guanosine tetra- and pentaphosphate) can trigger or inhibit cell growth and division, ensuring these life cycle phases occur in a coordinated manner. Eukaryotic growth and division also depend on nutrient availability and metabolic processes; however, there is an important difference from prokaryotes: metabolic signals trigger growth and division, but the reverse is also true, in that the life cycle can regulate metabolism. Proteins controlling cell growth and mitosis, such as cyclin-dependent kinases (CDKs), also regulate metabolism, while certain enzymes involved in carbohydrate metabolism control not only metabolic processes but also the transition between the different phases of the eukaryotic life cycle (Kalucka et al., 2015).

The second level of mechanistic integration is provided by various *regulatory proteins* that enable both prokaryotic and eukaryotic cells to grow and divide in an interdependent manner, as evidenced by their life cycles. These regulatory proteins facilitate the spatiotemporal coordination between developmental phases (such as cell size increase, DNA replication, transcription, and chromosome segregation) and cell division (including bacterial binary fission, eukaryotic mitosis, and meiosis). By "spatiotemporal coordination", I refer to the fact that growth and development phases occur not only in a sequential order, sometimes strictly determining other phases of the life cycle (as in eukaryotic cells), but also at specific sites within the intracellular space (e.g. the plasma membrane, cytoplasm, nucleoplasm). In bacteria, binary fission is spatiotemporally coordinated with DNA replication through regulatory proteins such as GidA and MioC, and with chromosome segregation via mechanisms like nucleoid occlusion, Ter linkage, and other protein complexes (e.g. MukBEF and MinCDE) (Dewachter et al., 2018). In eukaryotes, mitosis is coordinated with DNA replication (S phase), cell size growth, and protein synthesis (G1 and G2 phases) primarily through cyclin-dependent kinases and other regulatory proteins (e.g. the anaphase-promoting complex/cyclosome and the spindle assembly checkpoint).[22]

A third level of mechanistic integration is provided by *cytoskeletal proteins* that act as scaffolds, enabling not only chromosome segregation and septum formation but also crucial spatial coordination among different parts of the cell during division. This ensures that the cell reproduces as a cohesive whole. In bacteria, the FtsZ protein is fundamental, responsible for assembling the Z-ring, recruiting divisome components, constraining liposomes, and establishing the septum (Dewachter et al., 2018). Similarly, in eukaryotic cells, microtubules and their associated motor proteins (dynein and kinesins) not only shape the mitotic spindle and facilitate chromosome attachment but also promote the reshaping of the nuclear envelope, and the division and segregation of mitochondria, plastids, and endomembranes of the secretory pathway. This coordination ensures that the compartments of the eukaryotic cell divide and segregate in a manner that produces daughter cells morphologically similar to the parent (Bergeland et al., 2001; Wieser & Pines, 2015).

These three levels of mechanistic integration enable a cell to reproduce as a cohesive whole, thereby replicating its organisational structure in the daughter cells. Furthermore, because these levels coordinate cellular division with metabolic and developmental processes, reproduction is contingent on systemic conditions (such as nutrient availability, metabolic processes, energy production, cell size, and DNA replication), resulting in a *system-level coordinated reproduction*. The specific physiological (mechanistic) integration necessary to ensure the interdependence of metabolism, growth, and reproduction has profound implications for understanding biological autonomy and individuality.

As discussed extensively in Chap. 2, an organisational account of biological autonomy hinges on the concepts of constitutive and interactive autonomy. Constitutive autonomy involves processes such as metabolism, gene expression, and development that are essential for the self-maintenance of a biological system. Interactive autonomy refers to a living being's ability to actively engage with its environment, transforming and being transformed by it (Moreno & Mossio, 2015). Metabolism exemplifies constitutive autonomy by providing the energy necessary for a cell's self-maintenance. Similarly, growth and developmental processes are integral to constitutive autonomy, as the life cycle of an organism—birth, growth, reproduction, and death—demonstrates the ongoing process of self-maintenance.

From an organisational perspective, metabolism and growth are deeply interdependent, creating a system of organisational closure. Metabolism supplies the energy required to sustain both quiescent and growth states, while growth influences metabolism; regulatory proteins involved in development also govern metabolic processes, ensuring that growth dynamics shape metabolic activity. Thermodynamically, metabolism and development can be viewed as complementary and interdependent. Metabolism maintains a cell's state far from

[22] Interestingly, in prokaryotes, chromosome segregation is not part of the division cycle. Instead, in eukaryotes, chromosome segregation is part of the mitotic process occurring during anaphase.

thermodynamic equilibrium, temporarily contravening the second law of thermodynamics. Conversely, the end of an organism's life cycle, marked by death, represents the achievement of thermodynamic equilibrium, fulfilling the second law of thermodynamics.

The ontological status of reproduction is intricate, as it does not directly contribute to the self-maintenance of the current biological organisation (comprising metabolism and growth). Reproductive capacities are often termed "cross-generation functions" because they support the autonomous organisation of the lineage, species, or biological community, distinct from "intra-generation functions," which pertain to the self-maintenance of individual organisms (Saborido et al., 2011). Despite this distinction, both intra- and cross-generation functions are considered to share the same ontological status, as each contributes to the overall self-maintenance of biological systems. *Intra-generation functions*, including metabolic and developmental processes, sustain individual organisms on an *ontogenetic scale*. In contrast, *cross-generation functions* support the continuation of species across multiple generations on a *phylogenetic scale*.

Given this framework, the question arises as to whether intra- and cross-generation functions are physiologically connected and integrated. Based on the discussions presented in this chapter, I propose that there is a circular relationship between *metabolism*, *development*, and *reproduction*. Reproduction relies on metabolism to provide the energy necessary for division processes. Similarly, cellular reproduction is dependent on development, as division can only occur when an organism has reached a certain developmental stage. Conversely, metabolism and development are contingent upon reproduction because these processes unfold in organisms produced through reproductive mechanisms. Thus, reproduction generates new biological entities that maintain a material connection to their progenitors, ensuring that the intra-generation functions of the progeny are linked to the cross-generation functions of the reproducer.

The functional integration between intra- and cross-generation functions has significant implications for defining biological individuality, particularly in understanding the link between physiological and evolutionary individuality. As discussed in Chap. 5, physiological individuality refers to the extent to which a biological entity exhibits cohesive physiological integration, enabling it to function as a coordinated unit. Evolutionary individuality, on the other hand, encompasses the criteria necessary for natural selection as outlined by Lewontin (1970): genetic and phenotypic variation, differential fitness resulting from this variation, and the heritability of this variation.

In relation to physiological and evolutionary individuality, I offer two key observations: firstly, the integration of metabolism, growth, and system-level coordinated reproduction is central to *physiological individuality*; secondly, system-level coordinated reproduction is a prerequisite for *evolutionary individuality* (i.e. a unit of selection) and is dependent on physiological individuality, which in turn relies on the integration of metabolism, growth, and reproduction.

Regarding the first point, contemporary accounts of physiological individuality often emphasise dimensions such as metabolic (Dupré & O'Malley, 2009), immunological (Pradeu, 2010, 2016), and developmental (Griesemer, 2016) aspects of biological organisation. I propose that understanding physiological individuality requires acknowledging the role of the physiological integration between metabolism, development, and reproduction. These processes are fundamental to a cell's self-maintenance and thus to its physiology.[23] This understanding is underpinned by the three levels of mechanistic integration previously described: first, nutrient-dependent signals that link metabolic status to growth and indirectly to reproduction; second, regulatory proteins that integrate growth and division; and third, cytoskeletal and motor proteins that facilitate spatial coordination within the cell, ensuring system-level reproduction.

Regarding the second point, my perspective offers an advancement beyond current views of evolutionary individuality. Many existing theories emphasise the role of reproductive bottlenecks, where a single cell proliferates to form a multicellular organism (Maynard Smith & Szathmáry, 1995; Godfrey-Smith, 2009), and the distinction between germ and soma (Godfrey-Smith, 2009). However, these criteria have notable limitations: they are particularly well-suited to multicellular organisms, especially in the context of embryogenesis, but they fall short when applied to unicellular organisms, whether prokaryotic or eukaryotic. Additionally, these frameworks do not adequately address the biological connection between evolutionary individuality and the physiological processes that ensure self-maintenance.

I propose that for a unicellular organism to be considered an evolutionary individual, it must exhibit system-level coordinated reproduction, which is organisationally closed with metabolic and developmental processes and demonstrates the three levels of mechanistic integration outlined earlier. In essence, the concept of evolutionary individuality is intrinsically linked to physiological individuality. This perspective underscores that evolutionary individuality cannot be fully understood without considering the physiological mechanisms that underpin it.[24]

---

[23] The notion of physiology covers all those phenomena that allow a unicellular and multicellular organism to self-maintain in a steady state, keeping a few parameters (e.g. pH, temperature, intracellular calcium concentration, etc.) within certain "physiological" ranges. By contrast, pathophysiology occurs when several parameters are not within physiological ranges, thus progressively preventing the organism from self-maintaining.

[24] If it is true that evolutionary individuality entails physiological individuality, the reverse is not always true. As already pointed out by Pradeu (2016), there are physiological individuals that do not necessarily have a system-level collective reproduction. For example, in collective systems (e.g. colonies of bacteria or insects), there is no system-level coordinated reproduction but rather independent forms of reproduction of the parts of the whole (see also Sects. 5.3 and 5.5). Another case of physiological individuals that are not evolutionary units is represented by sexually reproducing organisms (through meiosis) that are sterile: despite having reproductive functions, they cannot perform them.

8.6 Concluding Remarks

In this chapter, I have examined the relationship between biological reproduction, functional integration, and heredity, addressing whether and how physiological integration is a necessary condition for system-level coordinated reproduction. This enables a biological organisation to transmit its genetic and phenotypic traits to offspring, thus leading to a parent-offspring lineage. To address this issue, I have analysed two representative examples of cellular reproduction: the binary fission of prokaryotes (bacteria and archaea) and the mitosis of eukaryotic cells. A comparative analysis of these two forms of reproduction reveals a common theoretical core: first, system-level coordinated reproduction is mutually dependent on developmental processes to generate a life cycle that both sustains and is sustained by metabolic processes. Secondly, this functional interdependence between system-level reproduction, growth, and metabolism relies on three levels of functionally integrated mechanisms: nutrient-dependent signals coordinating the life cycle with metabolism, regulatory proteins controlling the life cycle, and cytoskeletal and motor proteins controlling spatial coordination during cellular fission.

I have also observed that this organisational closure between reproduction, growth, and development sheds new light on issues of biological autonomy and individuality. I have emphasised that the integration between reproduction, growth, and development forms the physiological basis for unifying intra- and cross-generation functions and for understanding the continuity between constitutive processes and reproductive functions. Furthermore, I have stressed that the aforementioned organisational closure is the linchpin of physiological and evolutionary individuality, providing a unified view of what constitutes a biological individual.

I now address the third and final question of this chapter: Why did the proto-eukaryotic cell need to evolve the prokaryotic mechanism of binary fission into mitosis and meiosis? Answering this question is challenging due to the lack of fossils or living organisms that are intermediate between prokaryotes and eukaryotes. Nevertheless, I can suggest some theoretical clues.

First, the transition from prokaryotic to eukaryotic cells involved the emergence of many intracellular organelles, causing a significant change in the cytoplasmic space. For a proto-eukaryotic cell to divide and reproduce faithfully, it needed to develop reproductive mechanisms that allowed for the precise division and segregation of organelles to transmit them accurately to daughter cells. Since the prokaryotic protein FtsZ[25] does not form microtubules and cannot segregate organelles (Margolin, 2005), it seems likely that the appearance of microtubules and eukaryotic motor proteins was crucial for the development of mitosis. Without these proteins, the construction of the spindle, the division of the nucleus and organelles, and cytokinesis would not have been possible (Wickstead & Gull, 2011; Koumandou

[25] The FtsZ protein facilitates the fission of organelles of endosymbiotic origin, such as plastids and mitochondria (Margolin, 2005).

et al., 2013). Given that the faithful transmission of intracellular structures was necessary for the preservation and evolution of eukaryotic cells over time, it is reasonable to hypothesise that the appearance of eukaryotic cytoskeletal and motor proteins preceded or co-evolved with the emergence of intracellular membranes and the division between cytoplasm and nucleoplasm (Cavalier-Smith, 2010). In other words, mitosis likely emerged early in eukaryogenesis to meet the new physiological need for the faithful reproduction of the internal division of cytoplasmic space.

Secondly, the emergence of a new set of regulatory proteins, notably cyclin-dependent kinases (CDKs), was required for temporally coordinating mitosis with interphase. Despite being poorly understood, CDKs appear to be a de novo class of proteins that did not evolve from bacterial cell cycle regulators (Liu & Kipreos, 2000). The appearance of CDKs has been extremely important for coordinating energy and carbon metabolism with proliferation, as they can control metabolic fluxes directly and indirectly through phosphorylation of metabolic enzymes (Solaki & Ewald, 2018).

Thirdly, the transition from closed to open mitosis is closely connected with the increase in genome size of eukaryotic cells. Since nuclear volume scales linearly with genome size, the length of the mitotic spindle would have increased, making it unable to fit into the nucleus, thus necessitating the breakdown of the nuclear envelope (open mitosis) (Sazer et al., 2014, p. R1101). Although the transition from closed to open mitosis is not fully understood, it is possible that transposable elements[26] facilitated this change by enabling the mitotic spindle to access the pool of cytoplasmic microtubule protein (Pickett-Heaps, 1974).

Finally, the transition from mitosis to meiosis is complex, but Wilkins and Holliday (2009) propose a plausible hypothesis. They suggest that meiosis evolved from mitosis through changes in the underlying mechanisms of the mitotic cycle, given the close correspondence between meiotic and mitotic stages.[27]

References

Bergeland, T., Widerberg, J., Bakke, O., & Nordeng, T. W. (2001). Mitotic partitioning of endosomes and lysosomes. *Current Biology, 11*(9), 644–651. https://doi.org/10.1016/s0960-9822(01)00177-4

Boettcher, B., & Barral, Y. (2013). The cell biology of open and closed mitosis. *Nucleus, 4*(3), 160–165. https://doi.org/10.4161/nucl.24676

Boucrot, E., & Kirchhausen, T. (2007). Endosomal recycling controls plasma membrane area during mitosis. *Proceedings of the National Academy of Sciences of the United States of America, 104*(19), 7939–7944. https://doi.org/10.1073/pnas.0702511104

[26] Transposable elements are DNA fragments capable of changing their position within a genome, potentially causing genetic mutations that alter a cell's genetic identity and genome size.

[27] According to Wilkins and Holliday (2009), meiosis arose from mitosis through a single step: the insertion of homolog synapsis (i.e. the pairing of two homologous chromosomes during meiosis) into the mitotic cycle.

Carpenter, M. L., Assaf, Z. J., Gourguechon, S., & Zacheus Cande, W. (2012). Nuclear inheritance and genetic exchange without meiosis in the Binucleate parasite Giardia intestinalis. *Journal of Cell Science, 125*(10), 2523–2532. https://doi.org/10.1242/jcs.103879

Cavalier-Smith, T. (2010). Origin of the cell nucleus, mitosis and sex: Roles of intracellular coevolution. *Biology Direct, 5*, 7. https://doi.org/10.1186/1745-6150-5-7

D'Avino, P. P., Giansanti, M. G., & Petronczki, M. (2015). Cytokinesis in animal cells. *Cold Spring Harbor Perspectives in Biology, 7*(4), a015834. https://doi.org/10.1101/cshperspect.a015834

Dawkins, R. (1976). *The selfish gene*. Oxford University Press.

Dawkins, R. (1982). *The extended phenotype*. Oxford University Press.

Dewachter, L., Verstraeten, N., Fauvart, M., & Michiels, J. (2018). An integrative view of cell cycle control in Escherichia coli. *FEMS Microbiology Reviews, 42*(2), 116–136. https://doi.org/10.1093/femsre/fuy005

Difrisco, J., & Mossio, M. (2020). Diachronic identity in complex life cycles: An organizational perspective. In J. Dupré & A. S. Meincke (Eds.), *Biological identity: Perspectives from metaphysics and the philosophy of biology* (pp. 1–19). Routledge.

Dupré, J., & O'Malley, M. (2009). Varieties of living things: Life at the intersection of lineage and metabolism. *Philosophy, Theory, and Practice in Biology, 1*(3), e003. https://doi.org/10.3998/ptb.6959004.0001.003

Ereshefsky, M., & Pedroso, M. (2013). Biological individuality: The case of biofilms. *Biology & Philosophy, 28*(2), 331–349. https://doi.org/10.1007/s10539-012-9340-4

Ereshefsky, M., & Pedroso, M. (2015). Rethinking evolutionary individuality. *Proceedings of the National Academy of Sciences of the United States of America, 112*(33), 10126–10132. https://doi.org/10.1073/pnas.1421377112

Ewald, J. C., Kuehne, A., Zamboni, N., & Skotheim, J. M. (2016). The yeast cyclin-dependent kinase routes carbon fluxes to fuel cell cycle progression. *Molecular Cell, 62*(4), 532–545. https://doi.org/10.1016/j.molcel.2016.02.017

Fusco, G., & Minelli, A. (2019). *The biology of reproduction*. Cambridge University Press.

Gavet, O., & Pines, J. (2010). Activation of cyclin B1-Cdk1 synchronizes events in the nucleus and the cytoplasm at mitosis. *The Journal of Cell Biology, 189*(2), 247–259. https://doi.org/10.1083/jcb.200909144

Godfrey-Smith, P. (2009). *Darwinian populations and natural selection*. Oxford University Press.

Godfrey-Smith, P. (2013). Darwinian individuals. In F. Bouchard & P. Huneman (Eds.), *From groups to individuals: Evolution and emerging individuality* (pp. 17–36). MIT Press.

Griesemer, J. (2000). Development, culture, and the units of inheritance. *Philosophy of Science, 67*(S3), S348–S368. https://doi.org/10.1086/392831

Griesemer, J. (2016). Reproduction in complex life cycles: Toward a developmental reaction norms perspective. *Philosophy of Science, 83*(5), 803–815. https://doi.org/10.1086/687865

Hull, D. L. (1980). Individuality and selection. *Annual Review of Ecology, Evolution, and Systematics, 11*, 311–332. https://doi.org/10.1146/annurev.es.11.110180.001523

Imoto, Y., Yoshida, Y., Yagisawa, F., Kuroiwa, H., & Kuroiwa, T. (2011). The cell cycle, including the mitotic cycle and organelle division cycles, as revealed by cytological observations. *Journal of Electron Microscopy, 60*(Suppl 1), S117–S136. https://doi.org/10.1093/jmicro/dfr034

Kalucka, J., Missiaen, R., Georgiadou, M., Schoors, S., Lange, C., De Bock, K., Dewerchin, M., & Carmeliet, P. (2015). Metabolic control of the cell cycle. *Cell Cycle, 14*(21), 3379–3388. https://doi.org/10.1080/15384101.2015.1090068

Koumandou, V. L., Wickstead, B., Ginger, M. L., van der Giezen, M., Dacks, J. B., & Field, M. C. (2013). Molecular paleontology and complexity in the last eukaryotic common ancestor. *Critical Reviews in Biochemistry and Molecular Biology, 48*(4), 373–396. https://doi.org/10.3109/10409238.2013.821444

Lan, G., Daniels, B. R., Dobrowsky, T. M., Wirtz, D., & Sun, S. X. (2009). Condensation of FtsZ filaments can drive bacterial cell division. *Proceedings of the National Academy of Sciences, 106*(1), 121–126. https://doi.org/10.1073/pnas.0807963106

Lewontin, R. C. (1970). The units of selection. *Annual Review of Ecology, Evolution, and Systematics, 1*, 1–18. https://doi.org/10.1146/annurev.es.01.110170.000245
Liu, J., & Kipreos, E. T. (2000). Evolution of cyclin-dependent kinases (CDKs) and CDK-activating kinases (CAKs): Differential conservation of CAKs in yeast and metazoa. *Molecular Biology and Evolution, 17*(7), 1061–1074. https://doi.org/10.1093/oxfordjournals.molbev.a026387
Margolin, W. (2005). FtsZ and the division of prokaryotic cells and organelles. *Nature Reviews. Molecular Cell Biology, 6*(11), 862–871. https://doi.org/10.1038/nrm1745
McIntosh, J. R. (2016). Mitosis. *Cold Spring Harbor Perspectives in Biology, 8*(9), a023218. https://doi.org/10.1101/cshperspect.a023218
Militello, G., & Moreno, Á. (2018). Structural and organisational conditions for being a machine. *Biology & Philosophy, 33*(5), 35. https://doi.org/10.1007/s10539-018-9645-z
Moreno, A., & Mossio, M. (2015). *Biological autonomy: A philosophical and theoretical enquiry*. Springer.
Mossio, M., & Pontarotti, G. (2022). Conserving functions across generations: Heredity in light of biological organization. *The British Journal for the Philosophy of Science, 73*(1), 249–278. https://doi.org/10.1093/bjps/axz031
Ovejero, S., Bueno, A., & Sacristán, M. P. (2020). Working on genomic stability: From the S-phase to mitosis. *Genes, 11*(2), 225. https://doi.org/10.3390/genes11020225
Pickett-Heaps, J. (1974). The evolution of mitosis and the eukaryotic condition. *Biosystems, 6*(1), 37–48. https://doi.org/10.1016/0303-2647(74)90009-4
Pradeu, T. (2010). What is an organism? An immunological answer. *History and Philosophy of the Life Sciences, 32*(2–3), 247–267.
Pradeu, T. (2016). Organisms or biological individuals? Combining physiological and evolutionary individuality. *Biology and Philosophy, 31*(6), 797–817. https://doi.org/10.1007/s10539-016-9551-1
Saborido, C., Mossio, M., & Moreno, A. (2011). Biological organization and cross-generation functions. *British Journal for the Philosophy of Science, 62*(3), 583–606. https://doi.org/10.1093/bjps/axq034
Sazer, S., Lynch, M., & Needleman, D. (2014). Deciphering the evolutionary history of open and closed mitosis. *Current Biology, 24*(22), R1099–R1103. https://doi.org/10.1016/j.cub.2014.10.011
Smith, M., & Szathmáry, E. (1995). *The major transitions in evolution*. Oxford University Press.
Solaki, M., & Ewald, J. C. (2018). Fueling the cycle: CDKs in carbon and energy metabolism. *Frontiers in Cell and Developmental Biology, 6*, 93. https://doi.org/10.3389/fcell.2018.00093
Wallden, M., Fange, D., Lundius, E. G., Baltekin, Ö., & Elf, J. (2016). The synchronization of replication and division cycles in individual E. coli cells. *Cell, 166*(3), 729–739. https://doi.org/10.1016/j.cell.2016.06.052
Wang, J. D., & Levin, P. A. (2009). Metabolism, cell growth and the bacterial cell cycle. *Nature Reviews Microbiology, 7*(11), 822–827. https://doi.org/10.1038/nrmicro2202
Westfall, C. S., & Levin, P. A. (2017). Bacterial cell size: Multifactorial and multifaceted. *Annual Review of Microbiology, 71*(September), 499–517. https://doi.org/10.1146/annurev-micro-090816-093803
Wickstead, B., & Gull, K. (2011). The evolution of the cytoskeleton. *The Journal of Cell Biology, 194*(4), 513–525. https://doi.org/10.1083/jcb.201102065
Wieser, S., & Pines, J. (2015). The biochemistry of mitosis. *Cold Spring Harbor Perspectives in Biology, 7*(3), a015776. https://doi.org/10.1101/cshperspect.a015776
Wilkins, A. S., & Holliday, R. (2009). The evolution of meiosis from mitosis. *Genetics, 181*(1), 3–12. https://doi.org/10.1534/genetics.108.099762
Willis, L., & Huang, K. C. (2017). Sizing up the bacterial cell cycle. *Nature Reviews Microbiology, 15*(10), 606–620. https://doi.org/10.1038/nrmicro.2017.79

Chapter 9
Conclusions

Abstract This chapter synthesises the findings from the previous eight chapters to define functional integration. It introduces a broad concept of functional integration and critically assesses its explanatory value in understanding eukaryogenesis. Following this, it provides a more precise and robust characterisation of functional integration, exploring its theoretical implications within the ongoing philosophical debate on organismality, individuality, collective organisations, autonomy, major evolutionary transitions, and various forms of collective synchronic organisations. The chapter argues that functional integration arises from four biological dimensions: spatial constraints, system-level regulation, spatio-temporal coordination among the component parts, and system-level reproduction.

Keywords Functional integration · Spatial constraints · System-level regulation · Spatio-temporal coordination · System-level reproduction

In this final chapter of the book, I will connect the results from the eight chapters to draw conclusions about the nature and role of functional integration in one of the earliest evolutionary transitions: eukaryogenesis. I will begin by introducing a general concept of functional integration and critically evaluating its explanatory value for understanding eukaryogenesis. Subsequently, I will offer a more precise and robust characterisation of functional integration, considering its theoretical implications in the ongoing philosophical debate on organismality, individuality, collective organisations, autonomy, major evolutionary transitions, and various forms of collective synchronic organisations.

Based on the findings to date, a preliminary and tentative definition of functional integration can be framed as follows: the degree of functional interdependence among the component parts of a system that enables the whole to exhibit systemic behaviours. This formulation, consistent with current interpretations of functional integration (see Chap. 1), suggests that functional integration is a matter of *degree*: it occurs gradually and can vary in strength across different biological organisations. Consequently, functional integration results in qualitative differences in how

G. Militello, *Functional Integration*, History, Philosophy and Theory of the Life Sciences 38, https://doi.org/10.1007/978-3-031-94744-5_9

biological systems are functionally organised, as well as quantitative differences in their degree of integration.

Such a definition of functional integration allows us to place biological organisations on a spectrum of functional integration, evaluating their biological properties and behaviours accordingly. Different biological systems,[1] such as a molecular machine,[2] a single cell, and a bacterial colony, exhibit distinct kinds and degrees of functional integration. In a machine, functional integration depends on the cohesion of the parts, which enables them to function in a coordinated and organised manner to achieve a new and more complex function. In a single cell, functional integration refers to the degree of interdependence between cellular components that allows the cell to self-maintain and interact with its environment. In a collective organisation, functional integration involves the degree of cooperation and interdependence among individual components, which facilitates the emergence of systemic capacities and behaviours.

However, the aforementioned definition of functional integration captures only broad features of biological organisations and does not help us understand the origin of a markedly different and new form of integrated organisation, such as the eukaryotic cell. As discussed in Chaps. 4 and 5, a key event in *eukaryogenesis* was the engulfment of one or more prokaryotes by another and the subsequent transformation of the symbionts into organelles, resulting in a fully integrated individual. Therefore, the main challenge is to develop a concept of functional integration that can distinguish the qualitative differences between current symbiotic associations of prokaryotic cells and the organism-like organisation of a eukaryotic cell, which arises from a very specific and prolonged process of prokaryotic association. If we regard functional integration merely as a functional interdependence of parts that evolves in stages, we may fail to appreciate the structural and functional differences between a symbiotic association (e.g. a biofilm) and an organism-like organisation (e.g. the endosymbiosis between *Tremblaya* and *Moranella*). In both cases, the component parts (i.e. bacteria) exhibit functional interdependence, as they share

[1] Here, I use the term "biological system" in a very broad sense to denote a set of biological parts (or components) that perform one or more functions. Consequently, the term "biological system" can refer to the parts of cells (such as molecular machines), the cells themselves, or a group of cells working together (such as colonies).

[2] A clarification is necessary here. As noted in Militello and Moreno (2018), I am not suggesting that molecular machines are organisms, nor that they exhibit the same level of functional integration as a cell. What I assert is that, if we adopt a very general (and somewhat loose) definition of functional integration as the functional interdependence of parts performing a collective behaviour or function, it is reasonable to consider a molecular machine as a functionally integrated system. This is because a molecular machine performs its function due to the specific arrangement and integration of its parts. Furthermore, as highlighted by Militello and Moreno (2018), a molecular machine operates and performs its function through its functional interdependence with other macromolecules within the cellular context. However, if I provide a stricter and more rigorous definition of functional integration (as I will do later in this chapter), it becomes clear that a molecular machine is not functionally integrated in the same way as a cell or a colony of cells.

metabolic pathways, have common developmental processes, and respond to external stimuli in a highly coordinated manner.

We therefore require a more robust and specific concept of functional integration to differentiate between collective individuals and new forms of cohesive individuals. This entails refining the concept of functional integration in light of four specific organisational aspects identified through the case studies:

1. Functional integration arises from specific *spatial constraints*.
2. Functional integration results from *system-level regulation*.
3. Functional integration is the product of *spatio-temporal coordination* among the component parts.
4. Functional integration is a consequence of *system-level reproduction*.

Regarding the first point, Chaps. 2 and 5 highlight that the architecture of a system imposes significant constraints on how its parts function. In biological organisations, boundaries (e.g. membranes) act as spatial constraints on the area they enclose, a concept initially emphasised by Maturana and Varela (1973, 1980). Chapter 5 illustrated how the extracellular matrix and engulfment impose different spatial constraints on the behaviour of component parts, resulting in distinct forms of functional integration among symbionts. The extracellular matrix provides biofilm bacteria with global cohesion that enhances metabolic exchanges, intercellular communication, common development, and immune responses, making them more robust compared to their free-living counterparts. Nevertheless, biofilm bacteria retain a degree of autonomy that prevents most biofilms from exhibiting critical system-level capacities such as reproduction. Furthermore, the extracellular matrix is highly sensitive to environmental changes, leading to a somewhat transient, though effective, organisation.

In contrast, engulfment offers a common boundary that selectively constrains the behaviour of two bacteria, facilitating a strong integration between them. This process promotes extensive gene transfer from the endosymbiont to the host, allowing the host to progressively exert complete genetic control over the symbiotic association. Additionally, gene transfer is accompanied by efficient transport and targeting of functional components between the partners through the endosymbionts' membranes. As a result, the endosymbionts not only share metabolic pathways but also exhibit system-level reproductive capacities that enable the collective association to persist across generations and potentially evolve. For this reason, as suggested in Chap. 5, the engulfment of one prokaryote by another is a fundamental requirement for the emergence of a strong, non-facultative form of functional integration between different symbiotic partners.

Spatial constraints contribute to determining system-level regulation, which establishes functional control of the whole over its parts by responding to cues and signals. As highlighted in Chap. 5, the extracellular matrix of a biofilm and the engulfment process differ significantly in terms of system-level regulation. Despite exhibiting many coordinated physiological processes—such as shared metabolic pathways, collective developmental processes, or common immune responses—biofilms lack a collective genetic control system for their regulation. Consequently, biofilm bacteria maintain a degree of autonomy in performing their functions and exhibit distributed forms of control, without achieving a finely-tuned system-level control over development and reproduction. Similarly, the consortium of *Tremblaya* and *Moranella* does not display a collective genetic control, most likely because it represents an intermediate form between a symbiotic relationship and a new organism. Nevertheless, the endosymbiotic relationship has led to significant genomic reduction in both organisms, resulting in *Tremblaya* controlling some functions performed by *Moranella* and vice versa. I have referred to this unique situation as "interlocked regulation" (see Chap. 5), which could be considered a very primitive form of system-level regulation.

An important aspect of system-level regulation in symbiotic associations, as emphasised in Chap. 7, is the control of the sensorimotor capacities of individual components. In biofilms, the motility of bacteria is temporarily inhibited to support the formation and maintenance of the biofilm as a whole. In contrast, the endosymbiotic events that led to the eukaryotic cell involved the progressive loss of the sensorimotor abilities of endosymbionts and the control of their movement by the eukaryotic cytoskeleton. The loss of autonomous movement (or interactive autonomy) of organelles, in favour of the autonomy of the overall organisation, was a major step in eukaryogenesis and represents a significant difference from many symbiotic associations where component parts still retain the genes necessary for producing proteins involved in sensorimotor capacities and interactive autonomy.

Another fundamental aspect of functional integration, which also results from spatial constraints, is the establishment of effective communication (through signal exchange) among constituent parts to spatially and temporally coordinate their functions. Here, we can observe a significant difference between symbiotic associations and organism-like organisations. As demonstrated in Chap. 5, the extracellular matrix keeps bacteria closely together to enhance signal exchange among them. However, this signal exchange is not specific (i.e. it does not rely on individual interactions for specific effects) and can lead to considerable variability in behaviours within the collective system. In contrast, engulfment facilitates the development of precise mechanisms for transporting proteins across the endosymbiont's membrane and promotes the synthesis of complex regulatory macromolecules that act on the endosymbiont's membrane to ensure that the correct components are delivered to the right place at the right time.

In other words, engulfment enabled a high degree of spatio-temporal coordination between the component parts of a new individual. As highlighted in Chap. 6, the emergence of the eukaryotic cell involved the division of cytoplasmic space through intracellular membranes (endomembranes), which promoted spatial separation of the functions performed by the cell. For this organisation to be effective, not only systemic genetic regulation was required, but also a system of intracellular signalling that allows organelles to communicate with each other to coordinate their functions in space and time. I have shown that a major role in this spatiotemporal coordination of eukaryotic organelles is played by the cytoskeleton, which acts as a scaffold facilitating signal exchange between the organelles and the cytoplasm, thus enabling intracellular communication.

A fourth dimension of functional integration that I wish to emphasise in relation to the case studies is the achievement of system-level reproduction. The ability to reproduce as a whole is a fundamental characteristic of fully-fledged individuality (organisms) and is often absent in collective organisations, marking a potential dividing line between them. As discussed in Chap. 8, system-level reproduction can be accomplished through mechanisms that enable the overall system to divide and generate a new organisation with identical genetic and phenotypic features. For this to be viable, such mechanisms must establish a reciprocal causal loop with developmental and metabolic processes. The interdependence between reproduction, development, and metabolism is facilitated by cytoskeletal molecular machines, system-level regulation, and the spatio-temporal coordination of the parts.

My proposal for functional integration may offer new insights into significant issues in the current philosophical and biological debate. Firstly, clarifying functional integration can enhance our understanding of *organismality*. Current philosophical perspectives have rightly emphasised that functional integration is a fundamental aspect of organisms, characterised by the structural and functional cohesion of parts (Collier, 2004; Ruiz-Mirazo et al., 2000; Sober, 1991; Wolvekamp, 1966), a clear boundary between the organism and its environment (Godfrey-Smith, 2011; Sterelny & Griffiths, 1999), high levels of cooperation and low conflict among parts (Queller & Strassmann, 2009), and system-level reproduction (Godfrey-Smith, 2011; Okasha, 2011). My proposal aligns with these perspectives but places greater emphasis on the organisational aspects underlying the relationship between the organism as a whole and its constituent parts.

More precisely, structural and functional cohesion, as well as the distinction between interior and exterior, are not merely the result of spatial contiguity among the parts. Instead, they depend on a specific assembly of the parts (i.e. their architecture or design) and on a common boundary that acts as a spatial constraint on the behaviour of the parts. Queller and Strassmann's (2009) criterion of high cooperation and low conflict among the parts is somewhat general and does not explain how such cooperation can be achieved. In my view, low conflict among parts is facilitated by a system's ability to regulate its components systematically to prevent or resolve potential conflicts. High cooperation among parts is determined by effective communication and spatio-temporal coordination.

It has also been argued that system-level reproduction is a crucial aspect for defining the functional integration of an organism (Godfrey-Smith, 2011; Okasha, 2011). I concur with this view but emphasise that system-level reproduction should be considered alongside developmental processes (see Griesemer, 2000, 2016) and the metabolic processes that sustain it. In other words, the reciprocal loop between metabolism, development, and reproduction is a fundamental aspect of functional integration, thereby placing reproduction within a broader systemic physiological context.

For these reasons, I believe this proposal has significant implications for our understanding of biological individuality. The prevailing trend in the current philosophical debate is to adopt a pluralistic view of individuality, leading to what has been termed "promiscuous individualism" (Dupré, 2012). This pluralistic approach seeks to decompose the concept of "individual" into five main biological dimensions: structural unit, genetic unit, physiological unit, reproductive unit, and evolutionary unit (see Sects. 1.5.1 and 1.5.2). Pluralists argue that a broad and flexible concept of individuality is needed to encompass the diversity of biological entities.

In contrast, I propose a concept of functional integration that intersects with these five dimensions of individuality, offering a criterion to distinguish *cohesive individuals* (organism-like individuals) from the diverse array of *collective associations* (colonies, societies, etc.). My aim is not to determine whether a collective association qualifies as an individual but rather to provide a concept of functional integration that clarifies why and how an organism-like individual is more integrated than a collective association. This approach allows us to appreciate their qualitative differences. I assert that spatial constraints, system-level regulation, spatio-temporal coordination of the parts, and system-level reproduction are crucial parameters for defining and evaluating biological individuality.

Several authors (Dupré & O'Malley, 2009; Ereshefsky & Pedroso, 2013, 2015; Pradeu, 2016) have argued that symbiotic associations and colonies can be considered physiological individuals, as they share functional capacities (e.g. metabolism) and form cohesive physiological units. However, most symbiotic associations, colonies, and societies do not exhibit system-level reproduction and cannot produce parent-offspring lineages, raising questions about their status as levels of selection or evolutionary individuals (Clarke, 2016; Skillings, 2016). My proposal on functional integration is useful for addressing this issue: such organisations differ from organism-like individuals due to their lower internal functional integration, which typically manifests as weak spatial constraints, distributed (rather than hierarchical) regulation, lack of fine-tuned spatio-temporal coordination of the parts, and absence of system-level regulation. In essence, functional integration offers a standard for distinguishing between different biological organisations often referred to as "individuals".

A redefinition of functional integration also has implications for contemporary accounts of *biological autonomy* (Bickhard, 2000; Christensen, 2007; Collier, 2000; Kauffman, 2000; Moreno & Mossio, 2015; Rosslenbroich, 2014). In my view, biological autonomy is a characteristic of fully-fledged individuals and results from a high degree of functional integration that creates a reciprocal causal loop between the processes involved in the self-maintenance of the organisation (i.e. constitutive processes) and those used by the organism to functionally modify its environment (i.e. interactive processes, see also Arnellos et al. (2014)). Indeed, the autonomy of fully-fledged individuals (e.g. unicellular or multicellular organisms) involves a common (selective) boundary that constrains the parts, producing systemic regulation and spatio-temporal coordination not only over constitutive (including reproduction) processes but also over interactive ones.

The close relationship between functional integration and biological autonomy is also relevant for understanding collective organisations: these exhibit high functional integration among the individual living entities ("parts") at the expense of the whole. As a result, the individual parts retain their autonomy, and the overall synchronic organisation is prevented from being autonomous as a whole. In other words, the individual living entities that are part of such a collective organisation maintain their organismic condition.

A final theoretical issue that could benefit from my proposal on functional integration is the evolutionary and philosophical debate on *major evolutionary transitions*. According to some evolutionary biologists (Buss, 1987; Maynard Smith & Szathmáry, 1995; Michod, 1999), major evolutionary transitions are characterised by the emergence of new forms of cohesive individuals (e.g. unicellular eukaryotic organisms, multicellular organisms). These transitions result from a strong interdependence among the members of a group, leading them to "become so integrated that they evolve into new higher-level individuals" (Michod, 2007, p. 8613). This suggests that the previously mentioned aspects of functional integration help explain how a group of individuals can develop such a strong internal division of labour, high cooperation among the parts, and reduced conflict that it evolves from being merely a group of individuals into a new, fully-fledged, functionally integrated organism.

All in all, this work represents a preliminary theoretical effort to clarify and define the concept of functional integration. The central theme of this book is an exploration of the relationship between functional integration and biological individuality, which lies at the heart of any definition of an organism in both biology and the philosophy of biology. Despite its theoretical importance, relatively few biologists and philosophers have conducted in-depth analyses or developed a comprehensive framework for the concept of functional integration. This work aims to address this conceptual gap and is particularly valuable for illuminating critical issues at the intersection of theoretical biology and philosophy.

References

Arnellos, A., Moreno, A., & Ruiz-Mirazo, K. (2014). Organizational requirements for multicellular autonomy: Insights from a comparative case study. *Biology & Philosophy, 29*(6), 851–884. https://doi.org/10.1007/s10539-013-9387-x

Bickhard, M. H. (2000). Autonomy, function, and representation. *Communication and Cognition-Artificial Intelligence, 17*(3–4), 111–131.

Buss, L. W. (1987). *The evolution of individuality*. Princeton University Press.

Christensen, W. (2007). The evolutionary origins of volition. In D. Ross, D. Spurrett, H. Kincaid, & G. Lynn Stephens (Eds.), *Distributed cognition and the will: Individual volition and social context* (pp. 255–287). MIT Press.

Clarke, E. (2016). Levels of selection in biofilms: Multispecies biofilms are not evolutionary individuals. *Biology & Philosophy, 31*(2), 191–212. https://doi.org/10.1007/s10539-016-9517-3

Collier, J. D. (2000). Autonomy and process closure as the basis for functionality. *Annals of the New York Academy of Sciences, 901*, 280–290. https://doi.org/10.1111/j.1749-6632.2000.tb06287.x

Collier, J. D. (2004). Self-organization, individuation and identity. *Revue Internationale de Philosophie, 228*(2), 151–172. https://doi.org/10.3917/rip.228.0151

Dupré, J. (2012). *Processes of life: Essays in the philosophy of biology*. Oxford University Press. https://doi.org/10.1093/acprof:oso/9780199691982.001.0001

Dupré, J., & O'Malley, M. (2009). Varieties of living things: Life at the intersection of lineage and metabolism. *Philosophy, Theory, and Practice in Biology, 1*(3), e003. https://doi.org/10.3998/ptb.6959004.0001.003

Ereshefsky, M., & Pedroso, M. (2013). Biological individuality: The case of biofilms. *Biology & Philosophy, 28*(2), 331–349. https://doi.org/10.1007/s10539-012-9340-4

Ereshefsky, M., & Pedroso, M. (2015). Rethinking evolutionary individuality. *Proceedings of the National Academy of Sciences of the United States of America, 112*(33), 10126–10132. https://doi.org/10.1073/pnas.1421377112

Godfrey-Smith, P. (2011). Darwinian populations and transitions in individuality. In B. Calcott & K. Sterelny (Eds.), *The evolutionary transitions revisited*. MIT Press.

Griesemer, J. (2000). Development, culture, and the units of inheritance. *Philosophy of Science, 67*(S3), S348–S368. https://doi.org/10.1086/392831

Griesemer, J. (2016). Reproduction in complex life cycles: Toward a developmental reaction norms perspective. *Philosophy of Science, 83*(5), 803–815. https://doi.org/10.1086/687865

Kauffman, S. A. (2000). *Investigations*. Oxford University Press.

Maturana, H. R., & Varela, F. J. (1973). *De máquinas y seres vivos: una teoría sobre la organización biológia*. Editorial Universitaria.

Maturana, H. R., & Varela, F. J. (1980). *Autopoiesis and cognition: The realization of the living* (Boston studies in the philosophy and history of science) (Vol. 42). Springer. https://doi.org/10.1007/978-94-009-8947-4

Michod, R. E. (1999). *Darwinian dynamics: Evolutionary transitions in fitness and individuality*. Princeton University Press.

Michod, R. E. (2007). Evolution of individuality during the transition from unicellular to multicellular life. *Proceedings of the National Academy of Sciences of the United States of America, 104*(Suppl 1), 8613–8618. https://doi.org/10.1073/pnas.0701489104

Militello, G., & Moreno, Á. (2018). Structural and organisational conditions for being a machine. *Biology & Philosophy, 33*(5–6), 35. https://doi.org/10.1007/s10539-018-9645-z

Moreno, A., & Mossio, M. (2015). *Biological autonomy: A philosophical and theoretical enquiry*. Springer.

Okasha, S. (2011). Biological ontology and hierarchical organization: A defence of rank freedom. In B. Calcott & K. Sterelny (Eds.), *The major transitions in evolution revisited* (pp. 53–64). MIT Press.

Pradeu, T. (2016). Organisms or biological individuals? Combining physiological and evolutionary individuality. *Biology and Philosophy, 31*(6), 797–817. https://doi.org/10.1007/s10539-016-9551-1

Queller, D. C., & Strassmann, J. E. (2009). Beyond society: The evolution of organismality. *Philosophical Transactions of the Royal Society of London. Series B, Biological Sciences, 364*(1533), 3143–3155. https://doi.org/10.1098/rstb.2009.0095

Rosslenbroich, B. (2014). *On the origin of autonomy: A new look at the major transitions in evolution*. Springer.

Ruiz-Mirazo, K., Etxeberria, A., Moreno, A., & Ibáñez, J. (2000). Organisms and their place in biology. *Theory in Biosciences, 119*(3), 209–233. https://doi.org/10.1078/1431-7613-00018

Skillings, D. (2016). Holobionts and the ecology of organisms: Multi-species communities or integrated individuals? *Biology & Philosophy, 31*(6), 875–892. https://doi.org/10.1007/s10539-016-9544-0

Smith, M., & Szathmáry, E. (1995). *The major transitions in evolution*. Oxford University Press.

Sober, E. (1991). Organisms, individuals, and units of selection. In A. I. Tauber (Ed.), *Organism and the origins of self. Boston studies in the philosophy of science* (Vol. 129, pp. 275–296). Springer. https://doi.org/10.1007/978-94-011-3406-4_13

Sterelny, K., & Griffiths, P. E. (1999). *Sex and death. An introduction to philosophy of biology*. The University of Chicago Press.

Wolvekamp, H. P. (1966). The concept of the organism as an integrated whole. *Dialectica, 20*(2), 196–214. https://doi.org/10.1111/j.1746-8361.1966.tb00516.x

Index

G. Militello, *Functional Integration*, History, Philosophy and Theory of the Life Sciences 38, https://doi.org/10.1007/978-3-031-94744-5

The manufacturer's authorised representative in the EU is Springer Nature Customer Service Centre GmbH, Europaplatz 3, 69115 Heidelberg, Germany. If you have any concerns regarding our products, please contact ProductSafety@springernature.com

Printed and bound by CPI Group (UK) Ltd, Croydon, CR0 4YY
30/09/2026
02225572-0004